*Construction Project
Management*

Construction Project Management

Third Edition

Richard H. Clough
The University of New Mexico, Emeritus

Glenn A. Sears
Department of Civil Engineering
The University of New Mexico

A JOHN WILEY PUBLICATION
John Wiley & Sons, Inc.
NEW YORK / CHICHESTER / BRISBANE / TORONTO / SINGAPORE

Dedication: In loving memory of Jean L. Clough
and her many contributions to past editions.

Acknowledgments: Graphics by Mary Sears of Insignia Graphics.

In recognition of the importance of preserving what has been
written, it is a policy of John Wiley & Sons, Inc., to have books
of enduring value published in the United States printed on
acid-free paper, and we exert our best efforts to that end.

Library of Congress Cataloging-in-Publication Data:

Clough, Richard Hudson.
    Construction project management / Richard H. Clough, Glenn A.
Sears. --3rd ed.
        p.   cm.
    Includes index.
    ISBN 0-471-54608-9
    1.Construction industry--Management.   2. Industrial project
management.   I. Sears, Glenn A.   II. Title.
TH438.C62   1991
    624'.068--dc20
                                                        91-25488
                                                        CIP

Printed in the United States of America

10 9 8 7 6 5 4 3 2 1

Preface

The essential thrust of the third edition of *Construction Project Management* is to develop and discuss management methods that have been devised specifically for the direction and control of construction in the field. A construction project presents the contractor with many difficult and complex management problems. Because of the unique nature of construction, procedures and techniques are required that have been created especially for application to this important industry. This third edition is dedicated to the discussion and application of such methods.

The guidance and regulation of construction projects is now almost totally dependent upon the computer for the generation of detailed management data. Modern computer software is now available that produces a wide variety of management information that can be applied to the direction and control of field projects. Computer application is an indispensable ingredient to the conduct of a profitable construction enterprise. In this regard, however, the project manager must comprehend and apply the computer generated information to the daily conduct and control of field construction activities. To do this, the meaning and significance of the data produced must be thoroughly understood, a circumstance that requires a complete and basic comprehension of the procedures being applied and the methods being used to generate the management information. This third edition recognizes and stresses this fact. Consequently, emphasis in this text is placed not upon computer operation techniques or program characteristics, but giving the reader a thorough understanding of the basic methods and procedures involved and the practical significance of the data generated.

By design, this is a how-to-do-it treatment of construction project management. The approach to the subject matter is intended to be practical with emphasis placed on use and application. The management methods discussed herein are being successfully applied to the operations of contractors, construction managers, and others associated with the accomplishment of construction in the field. The management methods discussed herein have been thoroughly tested in practice. This text is specifically designed to acquaint the reader with a set of procedures that have been especially developed to accommodate the unique character of the construction industry. Stressing application and practical aspects, the management control of project time, cost, resources, and finance are thoroughly discussed. This revision can serve equally well as a classroom textbook or as a procedural guide for independent study. The level of presentation is such that the material covered is readily understandable by those who have only limited construction backgrounds. As a means of providing continuity and cohesiveness to the general topic of construction management, an Example Project consisting of several major segments is used throughout the book. Different parts of the Example Project are used for purposes of illustrating the subject matter.

For generations men, machines, materials and money have been the four M's of construction. Efficient usage of these four resources is the essence of construction management. However, in recent years, an important change is these basic resources has occurred. Women now constitute an important part of the construction industry. They occupy responsible positions in the field trades and at all levels of management. Terms like journeyman, foreman, and pile driverman have been used in the industry for literally hundreds of years. Because such words are the only ones generally recognized, these words are used in this text but are not meant to infer gender. At times, the word he or him is used as a singular pronoun for foremen, project manager, superintendent, or design professional. Such use of the masculine gender is done solely for the sake of readability and has no inference of gender. The authors of this text recognize the important contribution that women have made and are making to the construction industry.

The generous reception accorded the previous editions of this book have been most gratifying. In this third edition, a genuine effort has been make to present and discuss the latest project management procedures in a manner designed to best serve the needs of those who use the book. The construction industry, this nation's largest single productive effort, has an insatiable need for personnel who possess training and talent in the management area. We trust that this third edition will continue to assist with the education of such personnel.

Richard H. Clough
Glenn A. Sears

Albuquerque, New Mexico
April 1991

Contents

APPENDIXES

1

CONSTRUCTION PRACTICES

1.1 INTRODUCTION

The objective of this book is to present and discuss the management of construction projects during their accomplishment in the field. Such projects involve much time and expense, and close management control of them is required if they are to be completed within the established time and cost limitations. Developed and discussed are management techniques directed toward the control of cost, time, resources, and project finance during the field construction process. Emphasis is placed on practical and applied procedures of proven efficacy.

Effective management of a project requires a considerable background of general knowledge about the construction industry. The purpose of this first chapter is to familiarize the reader with certain fundamentals of construction practice that will be useful for a complete understanding of the discussions presented in later chapters.

1.2 THE CONSTRUCTION INDUSTRY

In terms of the dollar value of product produced, the construction industry is the largest single production activity of the American economy. The annual expenditure for total construction accounts for approximately 10 percent of the gross national product (GNP). Thus about one of every 10 dollars spent for goods and services in this country is a construction dollar. The construction industry is one of our nation's largest employers, being directly responsible for approximately 6 percent of the private employment (or 5 percent of the total jobs) in the United States.

The construction industry plays a basic role on the national scene. Not only does it touch the lives of virtually everyone on a daily basis, it occupies a fundamental position in the national economy. This large and pervasive industry is regarded as the bellwether industry of this country. Periods of national prosperity are usually associated with high levels of construction activity. One is the natural result of the other.

The construction industry is heterogeneous and enormously complex. There are several major classifications of construction that differ markedly from each other: housing, nonresidential building, heavy, highway, utility, and industrial. In addition, these construction types are further divided into many specialties such as electrical, concrete, excavation, piping, and roofing.

Construction work is accomplished by contractors who vary widely in size and work type performed. General contractors assume broad responsibility for a comprehensive work package, while specialty contractors (often subcontractors to the general contractor) limit their endeavors to a particular aspect of the project. Some very large contracting firms put more than 10 billion dollars worth of construction into place each year. The annual budgets of the largest construction firms rival the gross national products of many small countries. However, in terms of numbers of contracting companies, the construction industry is typified by small businesses.

1.3   THE CONSTRUCTION PROJECT

Construction projects are intricate and time-consuming undertakings. The total development of a project normally consists of several phases requiring a diverse range of specialized services. In progressing from initial planning to project completion, the typical job passes through successive and distinct stages that demand inputs from such disparate directions as financial organizations, governmental agencies, engineers, architects, lawyers, insurance and surety companies, contractors, and building tradesmen.

During the construction process itself, even a structure of modest proportions involves many skills, materials, and literally hundreds of different operations. The assembly process must follow a natural order of events that, in total combination, constitutes a complicated pattern of individual time requirements and restrictive sequential relationships among the many segments of the structure.

To some degree each construction project is unique, and no two jobs are ever quite alike. In its specifics, each structure is tailored to suit its environment, arranged to perform its own particular function, and designed to reflect personal tastes and preferences. The vagaries of the construction site and the possibilities for creative and utilitarian variation of even the most standardized building product combine to make each construction project a new and different experience. The contractor sets up its "factory" on the site and, to a large extent, custom builds each job.

The construction process is subject to the influence of highly variable and sometimes unpredictable factors. The construction team, which includes architects, engineers, building tradesmen, subcontractors, material dealers, and others, changes from one job to the next. All the complexities inherent to different construction sites such as subsoil conditions, surface topography, weather, transportation, material supply, utilities and services, local subcontractors, and labor conditions are an innate part of construction.

As a consequence of these circumstances, construction projects are typified by their complexity and diversity and by the nonstandardized nature of their production. The use of factory-made modular units may diminish this individuality somewhat, but it is unlikely that field construction will ever be able to adapt itself completely to the standardized methods and product uniformity of assembly-line production.

1.4 PROJECT STAGES

The genesis of a construction project proceeds in rather definite order with the following stages of development being typical.

A. Planning And Definition

Once the owner has identified a need for a new facility, it must define the requirements and delineate the budgetary constraints. Project definition involves establishing broad project characteristics such as location, performance criteria, size, configuration, layout, equipment, services, and other owner requirements needed to establish the general aspects of the project. Conceptual planning stops short of detailed design although a considerable amount of preliminary architectural or engineering work may be required. The definition of the work is basically the responsibility of the owner, although a design professional may be called in to provide technical assistance and advice.

B. Design

This phase involves the architectural and engineering design of the entire project. It culminates with the preparation of final working drawings and specifications for the total construction program. In practice, the design, procurement, and construction often overlap, with procurement and construction proceeding as various stages of the design are completed and drawings and specifications become available.

C. Procurement And Construction

Procurement refers to the ordering, expediting, and delivering of key project equipment and materials, especially those that may involve long delivery periods. This function may or may not be handled separately from the construction process itself. Construction is, of course, the process of physically erecting the project and putting the materials and equipment into place. This involves providing the manpower, construction equipment, materials, supplies, and supervision necessary to accomplish the work.

1.5   THE OWNER

The owner, whether public or private, is the instigating party that gets the project financed, designed, and built. Public owners are public bodies of some kind ranging from the federal government down through state, county, and municipal entities to a multiplicity of local boards, commissions, and authorities. Public projects are paid for by appropriations, bonds, or other forms of financing and are built to perform a defined public function. Public owners must proceed in accordance with applicable statues and administrative directives pertaining to advertising for bids, bidding procedure, construction contracts, contract administration, and other matters relating to the design and construction process.

Private owners may be individuals, partnerships, corporations, or various combinations thereof. Most private owners have the structure built for their own use: business, habitation, or otherwise. However, many private owners do not intend to become the end users. The completed structure is to be sold, leased, or rented to others. These parties may or may not be known to the owners at the time of construction.

1.6   THE ARCHITECT–ENGINEER

The architect–engineer, also known as the design professional, is the party or firm that designs the project. Because such design is architectural or engineering in nature, or often a combination of both, the term "architect–engineer" is used in this book to refer to the design professional, regardless of the applicable specialty or the relationship between the architect-engineer and the owner.

The architect–engineer can occupy a variety of positions with respect to the owner for whom the design is done. Many public agencies and large corporate owners maintain their own in-house design capability. In such instances, the architect–engineer is the design arm of the owner. The traditional and most common arrangement is where the architect–engineer is a private and independent design firm that accomplishes the design under contract with the owner. Where the design–construct mode of construction is used, the owner contracts with a single party for both design and construction. In such a case, the architect–engineer is a branch of or is affiliated in some way with the construction contractor.

1.7   THE PRIME CONTRACTOR

A prime contractor, also known as general contractor, is the firm that is in contract with the owner for the construction of a project, either in its entirety or for some specialized portion thereof. In this regard, the owner may choose to use a single prime contract or several separate prime contracts.

Under the single-contract system, the owner awards construction of the entire project to one prime contractor. In this situation, the contractor brings together all the diverse elements and inputs of the construction process into a single, coordinated effort and assumes full, centralized responsibility for the delivery of the finished job constructed in accordance with the contract documents. The general contractor is fully responsible to the owner for the performance of the subcontractors and other third parties to the construction contract.

When separate contracts are used, the project is not constructed under the centralized control of a single prime contractor. Rather, several independent contractors work on the project simultaneously, each being responsible for a designated portion of the work. Each of the contractors is in contract with the owner and each functions independently of the others. Hence each of these contractors is a prime contractor. Coordination of these contractors may be made the responsibility of the owner, the architect–engineer, a construction manager, or one of the prime contractors who is paid to perform certain overall job-management duties.

1.8   COMPETITIVE BIDDING

A prime contractor is selected by the owner on the basis of competitive bidding, negotiation, or some combination of the two. A major proportion of the construction in the United States is done by contractors who obtain their work in bidding competition with other contractors. The competitive bidding of public projects is normally required by law and is standard procedure for public agencies. Essentially all public construction work is done by this method. When bidding a project, the contractor estimates how much the structure will cost using the architect–engineer's drawings and specifications as a basis for the calculations. To this cost it adds what seems to be a reasonable profit and guarantees to do the entire job for the stated price.

Price amounts quoted by the bidding contractors most often constitute the principal basis for selection of the successful contractor, the low bidder usually receiving the contract award. Most bidding documents stipulate that the work shall be awarded to the "lowest responsible bidder." This gives the owner the right to reject the proposal of a bidding contractor if the contractor is judged to be unqualified for some reason. If its bid is selected, the contractor must complete the work in exchange for the contract amount.

Competitive bidding can also be used where the successful contractor is determined on a basis other than the estimated cost of the construction. Where the contract will involve the payment of a prescribed fee to the contractor, the amount of the fee is sometimes used as a basis of competition among contractors. To illustrate, construction management services are sometimes obtained by an owner using the fees proposed by the different bidders as the basis for contract award.

1.9   NEGOTIATED CONTRACTS

There are times when it can be advantageous for an owner to negotiate a contract for its project with a preselected contractor or small group of contractors. It is common practice for an owner to forgo the competitive bidding process and to handpick a contractor on the basis of its reputation and overall qualifications to do the job. A contract is negotiated between the owner and the chosen contractor. Such contracts can obviously include any terms and provisions that are mutually agreeable to the parties. Most negotiated contracts are of the cost-plus-fee type, a subject that will be developed more fully later. Normally this type of contract is limited to privately financed work because competitive bidding is a usual legal requirement for public projects except under extraordinary or unusual circumstances such as time of war, natural disaster, or other emergency.

1.10 COMBINED BIDDING AND NEGOTIATION

An owner will sometimes combine elements of both competitive bidding and negotiation. One such process is to have a bidding where the competing contractors are required to submit their qualifications, along with their bids, and are encouraged to tender suggestions as to how the cost of the project could be reduced. The owner then interviews those contractors whose proposals appear most favorable and negotiates a contract with one of them.

1.11 SUBCONTRACTING

The extent to which a general contractor will subcontract work depends greatly on the nature of the project and the contractor's own organization. There are instances where the job is entirely subcontracted, the general contractor providing only supervision, job coordination, and perhaps general site services. At the other end of the spectrum, there are those projects where the general contractor does no subcontracting, choosing to do the entire work with its own forces. In the usual case, however, the prime contractor will perform the basic project operations and will subcontract the remainder to various specialty contractors. Work of a type with which it is not experienced or for which it is not properly equipped is usually subcontracted. Qualified subcontractors are usually able to perform their construction specialty more quickly and at lesser cost than can the general contractor. In addition, many construction specialties have specific licensing, bonding, and insurance requirements.

When the prime contractor engages a specialty firm to accomplish a particular portion of the project, the two parties enter into a contract called a subcontract. No contractual relationship is thereby established between the owner and the subcontractor. When a general contractor sublets a portion of its work to a subcontractor, the prime contractor remains responsible under its contract with the owner for any negligent or faulty performance by the subcontractor. The prime contractor assumes complete responsibility to the owner for the direction and accomplishment of the total work. An important part of this responsibility is the coordination and supervision of the subcontractors.

1.12 DESIGN-THEN-CONSTRUCT

Traditionally, field construction has not been started until the architect–engineer has completed and finalized the design. This sequence is still predominant in the industry and is referred to as the design-then-construct procedure. However, while the singe-file completion of one step before the other is initiated is acceptable to owners on some projects, it is not acceptable to others. A number of financial considerations dictate the earliest possible completion date for many construction projects. It is possible to reduce the total design–construction time required for some projects by starting the construction before the total design has been accomplished.

1.13 FAST TRACKING

Fast tracking refers to the overlapping accomplishment of project design and construction. As the design of progressive phases of the work are finalized, these work packages are put under contract, a process commonly referred to as "phased construction." Early phases of the project are commensurately under construction while later stages are still on the drawing boards. This procedure of overlapping the design and construction can appreciably reduce the total time required to achieve project completion. For obvious reasons, fast tracking and phased construction can sometimes offer attractive advantages to the owner, and likewise can be the source of severe coordination problems.

1.14 CONSTRUCTION CONTRACT SERVICES

A myriad of contract forms and types are available to the owner for accomplishing its construction needs. All call for defined services to be provided under contract to the owner. The scope and nature of such services can be made to include almost anything the owner wishes. The selection of the proper contract form appropriate to the situation is an important decision for the owner and is one deserving of careful consideration and consultation.

   The construction contract can be made to include construction, design–construct, or construction management services. These concepts are discussed in the following three sections.

1.15 CONSTRUCTION SERVICES

A large proportion of construction contracts provides that the general contractor has responsibility to the owner only for the accomplishment of the field construction. Under such an arrangement, the contractor is completely removed from the design process and has no input into it. Its obligation to the owner is limited to constructing the project in full accordance with the contract terms.

Where the contractor provides construction services only, the usual arrangement is for a private architect–engineer firm to perform the design in contract with the owner. Under this arrangement, the design professional acts essentially as an independent contractor during the design phase and as an agent of the owner during construction operations. The architect–engineer acts as a professional intermediary between the owner and contractor and often represents the owner in matters of construction contract administration. Under such contractual arrangements, the owner, architect–engineer, and contractor play narrowly defined roles, and the contractor is basically in an adversary relationship with the other two.

1.16 DESIGN–CONSTRUCT

When the owner contracts with a single firm for both design and construction and possibly procurement services, this is referred to as a "design–construct" project. This form of contract is usually negotiated although, occasionally, it is competitively bid. Usually the contractor has its own design section with architects and engineers as company employees. In other cases, however, the architect–engineer can be a contractor's corporate affiliate or subsidiary, or the contractor can joint-venture with an independent architect–engineer firm for a given project or contract.

Basic to design–construct is the team concept. The owner, designer, and builder work cooperatively in the total development of the project. The contractor provides substantial input into the design process on matters pertaining to materials, construction methods, cost estimates, and construction time schedules. Recent years have seen increasing acceptance and usage of this concept by owners, largely due to the economies of cost and time that can be realized by melding the two functions of design and construction. Injecting contractor experience and expertise into the design process offers the possibility of achieving cost savings for the owner. Because fast tracking is possible under a design–construct contract, the owner may well have the beneficial use of the structure considerably before it would have under the more traditional design-then-construct arrangement.

A "turn-key" contract is similar to a design–construct contract. The difference lies in the greater range of responsibilities that the contractor undertakes on behalf of the owner under a turn-key arrangement. For example, a turn-key contract will often include such services as land selection and acquisition, project financing, project equipage procurement, and leasing of the completed facility.

1.17 CONSTRUCTION MANAGEMENT

Construction management is a term that is applied to the providing of professional management services to the owner of a construction project with the objective of achieving high quality at minimum cost. Such services may encompass only a defined portion of the construction program such as field construction or it may include total project responsibility. The objective of this approach is to treat project planning, design, and construction as integrated tasks within a construction system. Where construction management is used, a nonadversarial team is created consisting of the owner, the construction manager, the architect–engineer, and the contractor. The project participants, by working together from project inception to project completion, attempt to serve the owner's best interests in optimum fashion. By striking a balance between construction cost, project quality, and completion schedule, the management team strives to produce for the owner a project of maximum value within the most economic time frame. Construction management does not include design or construction services per se, but involves management direction and control over defined design and construction activities.

Construction management services are performed for the owner for a stipulated fee by design firms, contractors, and professional construction managers. Such services range from merely coordinating contractors during the construction phase to broad scale responsibilities over project planning and design, project organization, design document review, construction scheduling, value engineering, field cost monitoring, and other management services. Selection of the construction manager by the owner is sometimes accomplished by competitive bidding using both fee and qualifications as bases for contract awards. In the usual instance, however, the construction management arrangement is considered to be a professional services contract and is negotiated. These contracts normally provide for a fixed fee plus reimbursement of management costs.

1.18 FIXED-SUM CONTRACT

A fixed-sum contract requires the contractor to complete a defined package of work in exchange for a sum of money fixed by the contract. If the actual cost of the work should exceed this figure, the contractor absorbs the loss. The owner is obligated to make only such total payment as is stipulated in the contractual agreement. A fixed-sum contract may be either lump sum or unit price.

With the lump-sum contract the contractor agrees to complete a stipulated package of work in exchange for a single, lump sum of money. Use of this form of contract is obviously limited to those construction projects where both the nature and quantity of each work type can be accurately and completely determined before the contract sum is set.

A unit-price contract requires the contractor to perform certain well-defined items of work in accordance with a schedule of fixed prices for each unit of work put into place. The total sum of money paid to the contractor for each work item is determined by multiplying the contract unit price by the number of units actually done on the job. The contractor is obligated to perform the quantities of work required in the field at the quoted unit prices, whether the final quantities are greater or less than those initially estimated by the architect–engineer. This is subject to any contract provision for redetermination of unit prices when substantial quantity variations occur. Unit-price contracts are especially useful on projects where the nature of the work is well defined, but the quantities of work cannot be accurately forecast in advance of construction.

1.19 COST-PLUS-FEE CONTRACTS

Cost-plus-fee contracts provide that the owner will reimburse the contractor for all construction costs and pay a fee for its services. How the contractor's fee is determined is stipulated in the contract, and a number of different procedures are used in this regard. Commonly used are provisions that the fee shall be a stipulated percentage of the total direct cost of construction or that the fee shall be a fixed sum. Incentive clauses are sometimes included that give the contractor an inducement to complete the job as efficiently and expeditiously as possible through the application of bonus and penalty variations to the contractor's basic fee. A guaranteed maximum cost is frequently included in cost-plus contracts. Under this form, the contractor agrees that it will construct the total project in full accordance with the contract documents and that the cost to the owner will not exceed some total upset price.

1.20 WORK BY FORCE ACCOUNT

The owner may elect to act as its own constructor rather than have the work done by a professional contractor. If the structure is being built for the owner's own use, this method of construction is called the force-account system. In such a situation, the owner may accomplish the work with its own forces, providing the supervision, materials, and equipment itself. On the other hand, the owner may choose to subcontract the entire project, assuming the responsibility of coordinating and supervising the work of the subcontractors. Because public projects must generally be contracted out on a competitive-bid basis, force-account work by a public agency is usually limited to maintenance, repair, or cases of emergency.

Many studies over the years have repeatedly revealed that most owners cannot perform field construction work nearly as well or as cheaply as professional contractors. The reason for these findings is obvious. The contractor is intimately familiar with materials, equipment, construction labor, and methods. It maintains a force of competent supervisors and workmen and is equipped to do the job. Only when the owner conducts a steady and appreciable volume of construction and applies the latest field management techniques is it economically feasible for it to carry out its own construction operations.

1.21 SPECULATIVE CONSTRUCTION

When owners, acting as their own general contractors, build structures for sale or lease to other parties, they engage in what is commonly referred to as speculative construction. Housing and commercial properties like shopping centers and warehousing facilities are common examples of such construction. In tract housing, for instance, "merchant" builders develop land and build housing for sale to the general public. This is a form of speculative construction where developers act as their own prime contractors. They build dwelling units on their own account and employ sales forces to market their products. In much speculative housing, contractors build for unknown buyers. In commercial applications, however, construction does not normally proceed until suitable sale or lease arrangements have been made. This is usually necessary so that the developers can arrange their financing as well as to enable them to build to the lessee's specifications. Most speculative builders function more as land or commercial developers than contractors, choosing to subcontract all or most of the actual construction work.

1.22 MANAGEMENT DURING THE DESIGN PHASE

Project cost and time control actually begin during the design phase. In the initial design stages, estimates such as annual cost to the owner and total life-cycle costs of the facility are made. Technical job standards are weighed against cost, function, maintenance, and appearance with the objective of minimizing the full cost of constructing, operating, and maintaining the new facilities over their useful life. As the design develops, construction method and material alternatives are subjected to value analysis as a rational means of optimizing the entire construction process in terms of cost and time. Cost budgets, ranging from preliminary to final, are prepared as the design approaches completion.

Time control during the design stage is directed toward minimizing construction time consistent with project quality and total cost. The delivery times of materials and project equipment are checked. Where long delivery periods are involved, the design is changed or procurement is initiated as soon as the design has progressed sufficiently to allow detailed purchasing specifications to be drawn. Construction methods are chosen whose cost characteristics are favorable and for which adequate labor and construction equipment will be available as needed. A preliminary project time schedule is usually prepared as the design progresses.

1.23 MANAGEMENT OF FIELD CONSTRUCTION

Discussions up to this point have demonstrated that owners have the option of using many different procedures to get their structures built. Regardless of the variability of these procedures, however, one party assumes management responsibility for the field construction process. Depending on the methods used by the owner, this party may be the owner, the architect–engineer, a construction manager, or a general contractor.

The management of field construction is customarily on an individual project basis, with a project manager being made responsible for all aspects of the construction. On large projects, a field office is usually established directly on the job site for the use of the project manager and his staff. Working relations with a variety of outside persons and organizations such as architects, engineers, owners, subcontractors, material and equipment dealers, labor unions, and regulatory agencies are an important part of guiding a job through to its conclusion. Field project management is directed toward pulling together all the diverse elements necessary to complete the project satisfactorily. Management procedures presented later will, in general, be discussed only as they apply to field construction.

1.24   THE PROJECT MANAGER

The project manager organizes, plans, schedules, and controls the field work and is responsible for getting the project completed within the time and cost limitations. He acts as the focal point for all facets of the project and brings together the efforts of all organizations having inputs into the construction process. He coordinates matters relevant to the project and expedites project operations by dealing directly with the individuals and organizations involved. In any such situation where events progress rapidly and decisions must be consistent and informed, the specific leadership of one person is needed. Because he has the overall responsibility, the project manager must have broad authority over all elements of the project. The nature of construction is such that he must often take action quickly on his own initiative, and it is necessary that he be empowered to do so. To be effective he must have full control of the job and be the one voice that speaks for the project. Project management is a function of executive leadership and provides the cohesive force that binds together the several diverse elements into a team effort for project completion.

Large projects normally will have a full-time project manager who is a member of the firm's top management or who reports to a senior executive of the company. The manager may have a project team to assist him or he may be supported by a central office functional group. When smaller contracts are involved, a single individual may act as project manager for several jobs simultaneously. An important aspect of a project manager's position is that his duties are normally separate from those of field supervision. The day-to-day direction of field operations is handled by a site supervisor or field superintendent. His duties involve working with the foremen, coordinating the subcontractors, directing construction operations, and keeping the work progressing smoothly and on schedule. The fact is that construction project authority is a partnership effort between the project manager and the field superintendent who work very closely together. Nevertheless, centralized authority is necessary for the proper conduct of a construction project, and the project manager is the central figure.

1.25   PROJECT MANAGER QUALIFICATIONS

To be effective, the project manager must possess three essential attributes. First, he must have a considerable background of practical construction experience so that he is thoroughly familiar with the workings and intricacies of the industry. Without such a basic grounding in construction fundamentals, the project manager would be completely unprepared to carry out his responsibilities.

Second, the project manager must have, or have available to him, persons with expertise and experience in the application of specialized management techniques to the planning, scheduling, and control of construction operations. These procedures have been developed specifically for application to construction projects and are those discussed herein. Because much of the management system is usually computer based, the project manager must have access to adequate computer support services.

Third, the project manager must have the personality and insight that will enable him to work harmoniously with other people, often under very strained and trying circumstances. The manager, after all, cannot accomplish everything through his efforts alone. He must work with and through people in the performance of his duties. This requires an appreciation and understanding of the human factor. Without this, his other attributes, however commendable, will be of limited effectiveness.

2

THE MANAGEMENT SYSTEM

2.1 NEED FOR PROJECT MANAGEMENT

If a project is to be constructed within its established budget and time schedule, close management control of field operations is a necessity. Project conditions such as technical complexity, importance of timely completion, resource limitations, and substantial costs put great emphasis on the planning, scheduling, and control of construction operations. Unfortunately, the construction process, once it is set into motion, is not a self-regulating mechanism and requires expert guidance if events are to conform to plans.

It must be remembered that projects are one-time and largely unique efforts of limited time duration which involve work of a nonstandardized and variable nature. Field construction work can be profoundly affected by events that are difficult, if not impossible, to anticipate. Under such uncertain and shifting conditions, field construction costs and time requirements are constantly changing and can seriously deteriorate with little or no advance warning. Skilled and unremitting management effort is not just desirable, it is absolutely imperative for a satisfactory final result.

Under most competitively bid, fixed-sum contracts calling for construction services only, the general contractor exercises management control over the construction operations. Self-interest is the essential motivation in such a case, the contractor being obligated by contract to meet a prescribed completion date and to finish the project for a stipulated sum. The surest way for the contractor to achieve its own objectives — and those of the owner in the bargain — is by applying some system of project management.

To serve the best interests of the owner is the primary emphasis of project control under other forms of contracts. Field management under design–construct, construction management, and many cost-plus contracts is principally directed toward providing the owner with professional advisory and management services to best achieve the owner's objectives.

2.2 PROJECT MANAGEMENT CHARACTERISTICS

In its most common context, the term "management" relates to the planning, organizing, directing, and controlling of a business enterprise. Business management is essentially a continuing and internal activity involving that company's own personnel, finances, property, and other resources. Construction project management, on the other hand, applies to a given project, the various phases of which are usually accomplished by different organizations. Therefore, the management of a construction project is not so much a process of managing the internal affairs of a single company as it is one of coordinating and regulating all of the inputs needed for the accomplishment of the job at hand. Thus, the typical project manager must work extensively with organizations other than his own. In such circumstances, much of his authority is conferred by contractual terms or power of agency and is, therefore, less direct than that of the usual business manager. Project management is largely accomplished through personnel of different employers working closely together.

2.3 DISCUSSION VIEWPOINT

It has been mentioned previously that the responsibility for field construction management is placed on different parties, depending on owner preference and the nature of the contracting procedure. Whether the owner, architect–engineer, general contractor, or a construction manager performs such duties is very much a matter of context. The basics of the pertinent management procedures are essentially the same, however, regardless of the implementing party. Nevertheless, to show detailed workings and examples of such management methods, it is necessary to present the material from the specific viewpoint of one of these parties. Thus, the treatment of management methods herein will be from the particular viewpoint of the general contractor where such designation is required by the nature of the discussion.

2.4 MANAGEMENT PROCEDURES

Field construction has little in common with the assembly-line production of standardized products. Standard costs, time-and-motion studies, process flowcharts, and line-of-balance techniques, traditional management devices used by the manufacturing industries have not lent themselves well to general construction application. Historically, construction project management has been a rudimentary and largely intuitive process, aided by the useful but inadequate bar chart (see Section 5.28.)

Over the years, however, new scientific management concepts have been developed and applied. Application of these principles to construction has resulted in the development of techniques for the management control of construction cost, time, resources, and project finance, treating the entire construction process as a unified system. Comprehensive management control is applied from inception to completion of construction operations.

Field project management starts with the onset of construction. At this point a comprehensive construction budget and a detailed time schedule of operations are prepared. These constitute the accepted cost and time goals used as a flight plan for the actual construction process. After the project has been started, monitoring systems are established that measure actual costs and progress of the work at periodic intervals. The reporting system provides progress information that is measured against the programmed targets. Comparison of field expense and progress with the established plan quickly detects exceptions that must receive prompt management attention. Data from the system can be used to make corrected forecasts of costs and time to complete the work.

The process just described is often called a "management-by-exception" procedure. When applied to a given project, it emphasizes the prompt and explicit identification of deviations from an established plan or norm. Reports that highlight exceptions from the standard enable the manager to recognize quickly those project areas requiring attention. So long as an item of work is progressing in accordance with plan, no action is needed. There are always plenty of problem areas that do require attention. Management-by-exception devices are useful, and this book emphasizes their application.

In addition to cost and time, the field management system is necessarily concerned with the management of job resources and with project financial control. Resources in this context refer to materials, labor, construction equipment, and subcontractors. Resource management is primarily a process of the advance recognition of project needs, scheduling and expediting the resources required, and leveling the demands where necessary. Project financial control involves the responsibility of the project manager for the total cash flow generated by the construction work and terms of the contract.

As indicated by the preceding discussion, there are several different aspects of a project control system. Each of these major management topics is treated separately in the chapters following. It must be recognized, however, that these aspects are highly interrelated segments of a total management process.

2.5   TIME AND COST MANAGEMENT

Project time and cost management are based on time and cost schedules developed for the project and an information system that will provide data for comparing expected with actual performance. The information or monitor system measures, evaluates, and reports job progress, comparing it with the performance planned. In this way, the project manager is apprised of the nature and extent of any deviation. When deviations do occur, the manager takes whatever action is considered to be feasible and effective to correct the situation. Costs and time can quickly get out of hand on construction projects where production conditions are so volatile. Job monitoring must quickly detect such aberrations. Cost and time control information must be timely with little delay between field work and management review of performance. This gives the project manager a chance to evaluate alternatives and take corrective action while an opportunity still exists to rectify problem areas.

In a sense, all management efforts are directed toward cost control because expedient project completion represents both construction savings for the contractor and beneficial usage for the owner. In practice, however, time and cost management are spoken of and applied as separate, although interrelated, procedures. One aspect of this separateness is the difference in job breakdown structure used for time and cost control purposes. The distinctive character of the two procedures requires that the project be divided into two different sets of elements: project components for time control and work classifications for cost control.

The realities of a field project make the strict control of every detail unattainable in a practical sense. Consequently, it must be recognized that the time and cost management methods discussed in this book are imperfect procedures, affording results of reasonable accuracy and timeliness to managers whose powers to control are far from absolute. Project management procedures offer no panacea for construction problems. They provide no magic answers, and the management information generated is no better than the quality of the input data. Nevertheless, a reasonably good basis is established for informed decision making.

2.6 PLANNING AND SCHEDULING

Planning, the first step in the process of construction time control, is discussed at length in Chapter 4. Planning establishes, on the basis of a detailed study of job requirements, what is to be done, how it is to be done, and the order in which it will proceed. The planning function is accomplished by dividing the project into many components or time-consuming steps, called "activities," and establishing the sequence in which they will be performed. An example of an activity might be "Install boiler" or "Set bar joists." The results of project planning are shown graphically in the form of a network diagram. This diagram can be drawn using either of two different graphical notation systems, "precedence" or "arrow." Precedence notation is emphasized herein and is used throughout for discussion purposes. The arrow convention is presented in Chapter 6.

A detailed time study of the planning network is then conducted, with adjustments to the plan being made as necessary to meet the project completion date. Some selective shortening of key construction activities may be in order at this point. Manpower and construction equipment requirements are evaluated for the individual job activities, with adjustments being made to minimize unbalanced or conflicting demands. On the basis of these studies the contractor establishes a calendar-date schedule of the anticipated start and finish times for each activity. The resulting time schedule, subject to periodic revision and correction during construction, is the essential basis for the day-to-day time control of the project. Such a schedule serves as an exceptionally effective early-warning device for detecting when and where the project is falling behind schedule. The several facets of project scheduling are the topics of Chapters 5, 7, and 8.

2.7 THE CPM PROCEDURE

The planning and scheduling of construction projects normally uses a network-based management procedure referred to as the Critical Path Method (CPM). CPM was developed especially to provide an effective and workable procedure for the planning and scheduling of construction operations. Widely used by the construction industry, and frequently a contract requirement, CPM involves a definite body of management procedures and is the basis for the planning and scheduling methods discussed in this book.

The heart of CPM is a graphical job plan that shows all the construction operations necessary for job completion and the order in which they will be done. This graphical network portrays, in simple and direct form, the complex time relationships and constraints among the various segments of a project. It has the tremendous advantage of easily accommodating modifications, refinements, and corrections. It provides the project manager with the following invaluable time-control information and devices:

1. Concise information regarding the planned sequence of construction operations.
2. A means to predict with reasonable accuracy the time required for overall project completion and the times to reach intermediate construction goals (commonly called "milestones").
3. Proposed calendar dates at which it is planned that the several activities of the project will be started and finished.
4. Identification of those "critical" activities whose expedient execution is crucial to timely project completion.
5. A guide for project shortening when the completion date must be advanced.
6. A basis for scheduling subcontractors and material deliveries to the job site.
7. A basis for balanced scheduling of manpower and construction equipment on the project.
8. The rapid evaluation of the time requirements of alternative construction methods.
9. A convenient vehicle for progress reporting, recording, and analysis.
10. A basis for evaluating the time effects of construction changes and delays.

2.8 TIME MONITORING AND CONTROL

When field operations begin, the order in which the project proceeds is in accordance with an approved job plan. During the construction period, advancement of the work is monitored by measuring and reporting the field progress at regular intervals. These data are analyzed and time-control measures are taken as appropriate to keep the work progressing on schedule.

Progress measurement for time-control purposes is an approximate process and is based on determining the time status of each individual job activity. Progress is normally measured by noting those activities that have been completed and estimating the times required to complete those in process. When compared with the latest planned schedule, these data give the manager an immediate indication of the time status of each job activity. Because activities seldom start or finish exactly as scheduled, the field information also serves as the basis for occasional updatings that yield revised project completion dates and corrected time schedules for the construction yet to be done. The workings of project time control are discussed in Chapter 9.

2.9 THE PROJECT COST SYSTEM

The project cost system is concerned with the control of expenses on current projects and the gathering of production information for use in estimating the cost of new work. The application of cost controls to a construction project actually begins when the costs are initially estimated. It is then that the project budget is established. This is the budget used by the project manager for cost-control purposes during field construction.

If there is to be an opportunity for genuine cost control, it must be possible to detect cost overruns promptly by making frequent comparisons between actual and budgeted expenses of production during the construction process. In addition, the actual costs must be determined in sufficient detail to enable project management to locate the trouble, should expense overruns occur. During the time the job is underway, cost accounting methods are applied to obtain the actual production rates and costs as they occur. Summary reports are prepared periodically that are especially designed to pinpoint work areas where costs are exceeding the budget. In this instance of "management-by-exception" the cost system immediately identifies for the project manager where production costs are unsatisfactory and management action is needed. By taking suitable corrective measures, the too-high costs can hopefully be reduced to more reasonable levels.

In addition to maintaining a continuous check on production costs for cost-control purposes, the project cost system yields valuable information needed for the estimating of future construction work. Average production rates and unit costs are obtained from completed projects and maintained in permanent files. These records of past experience are an invaluable resource to the estimator when new projects are being priced.

For both cost-control and estimating purposes, a construction project is divided into the same elementary work classifications or work types. Some examples of work types might be "footing concrete, place (cy)" or "structural steel, erect (ton)." These classifications are used throughout a company's cost system. Each work type is assigned a unique and permanent cost code number which is used consistently by all company personnel and which does not change from project to project. Chapter 10 presents a detailed treatment of a project cost system.

2.10   ESTIMATING THE PROJECT

When the project design has been finalized, a complete and detailed cost estimate is prepared. The contractor uses this estimate for bidding and subsequent cost-control purposes if a competitive-type contract is involved. With cost-plus and construction management contracts, a similar estimate is compiled essentially for the owner's cost-control purposes during construction. The final estimate is based on a detailed quantity takeoff that is a compilation of the total amounts of elementary work classifications required. The costs of labor, construction equipment, and materials are computed on the basis of the work quantities involved. Subcontract amounts are obtained from bids submitted by subcontractors to the general contractor. Taxes, overhead, and surety bonds are added as required.

Of all the costs involved in the construction process, those of labor and construction equipment are the most difficult to estimate and control. Fundamentally, the estimating of such costs is based on production rates. A production rate can be expressed as hours of labor or equipment time required to accomplish a unit amount of a given work type. An example of this is the number of labor hours required to erect a ton of structural steel. Production rates can also be expressed as the number of units of a work type that can be done per unit of time such as per hour. An instance of such a rate is the number of bank cubic yards of excavation that a power shovel can perform in one hour. For quick and convenient application, production rates are frequently converted to costs per unit of work. The source of production rates and unit costs is the company's cost-accounting system. When the cost estimate has been completed, the project control budget is prepared. This schedule of costs is the standard to which the actual costs of production are compared during field operations. Estimating project costs and preparing the project budget are discussed in Chapter 3.

2.11   PROJECT COST ACCOUNTING

Project cost accounting is the process of obtaining actual production rates and unit costs from ongoing projects. This system provides the basic information for project cost control and for estimating new work. Because of the uncertain nature of labor and equipment costs, these two items of expense are subjected to detailed and frequent analysis during the construction period. They are the main emphasis of a construction cost-accounting system. Basic inputs into the cost-accounting system are hours of labor and equipment time expended, hourly expense rates for labor and equipment, and quantities of work accomplished. These data are analyzed and periodic cost reports are produced. These summary reports compare budgeted with actual costs of production for cost-control purposes. These project cost reports not only enable comparisons to be made between budgeted and actual expenses as the work proceeds, but also provide a basis for making forecasts of the final project cost. In addition to cost reports of one kind or another, production rates and unit costs are generated for future use in estimating.

Cost accounting, unlike financial accounting, is not conducted entirely in terms of cost. To produce production rates and unit costs, work quantities and hours of labor and equipment usage must also be determined. Consequently, measuring and reporting work quantities put in place and hours of labor and equipment expended in the field are integral parts of a cost-accounting system.

2.12 RESOURCE MANAGEMENT

Resources refer to manpower, construction equipment, materials, and subcontractors. These resources totally control job progress and must be carefully managed during the construction process. Schedules of future resource needs are prepared and positive steps taken to assure that adequate job support will be available as required. Favorable material deliveries require skilled attention to procurement, shop-drawing checking and approval, expediting, and quality control. Labor crews and construction equipment must be scheduled and arranged. Subcontractors must be kept aware of the overall job schedule, given advance notice when their services are required, and their work must be coordinated into the total project effort.

Resource management involves other aspects as well. Job schedules must occasionally be adjusted to reduce the daily demand for certain resources to more practical levels. Impracticable bunching of job resources must be leveled to a smoother and more attainable time pattern of need. Resource management and its procedures are discussed in Chapter 8.

2.13 PROJECT FINANCIAL CONTROL

For management purposes a construction project is treated essentially as a separate and autonomous effort requiring resources and inputs from a variety of sources. For income and expense, profit or loss, and general financial accounting purposes, each project is handled separately and individually. The significance of this condition is that the project manager has responsibility for the control of project financial matters generally. Concerned here are considerations that range from total project cash flow to everyday matters of contract administration. Monthly pay requests, estimated schedule of payments by owner, project cash forecasts, changes to the contract, and disbursements to material dealers and subcontractors are all examples of project financial matters subject to management control procedures. Methods and procedures applicable to project financial management are discussed in Chapter 11.

2.14 PROJECT MANAGEMENT AND THE COMPUTER

The management and control of project time, cost, resources, and finance by the contractor during the field construction process require that the project manager originate, manipulate, summarize, and interpret large volumes of numerical data. In order to generate such information and apply it in optimum fashion, the project manager customarily relies on the computer to provide a wide range of computing and data generation services. Project managers must react quickly to changing conditions, and their decisions must be made with the secure knowledge that they are acting on the basis of adequate, accurate, and current intelligence. The computer can greatly assist in making this possible as well as providing information for the evaluation of alternate courses of action. With the computer doing most of the time-consuming data manipulation, the project manager can devote more time to problem solving and developing more profitable approaches.

However, because of the continuous emergence of new machines and new software support, no effort is made herein to discuss specific hardware configurations or the workings of computer programs that are in current usage for project management purposes. Rather, emphasis is placed on sources of input information, the management significance of the data generated by the computer, and how these data are applied to the control of a construction project.

2.15 MANUAL METHODS

The preceding discussion of computer application to job management is not meant to imply that manual methods have no place in the system. The project manager may rely on hand methods for limited portions of a project and to carry out computations for making quick checks, to determine the effect of changes, or to study a specialized portion of the work.

Even when the calculations are done by machine, the project manager must understand the computational procedures that are an innate part of the techniques applied. The manager's intimate familiarity with the workings of the procedures will provide an intuitive feel and grasp of a project that cannot be obtained in any other way. Because manual methods are useful in their own right and a thorough understanding of the computational methods involved with the computer generation of management data is crucial to the proper application of project control methods, this book discusses in step-by-step fashion the manual calculation of the several kinds of management information that are generated.

2.16 DISCUSSION FORMAT

In the remaining chapters, several different management procedures are presented. In an attempt to provide a sense of continuity in going from one topic to another, an Example Project is used as a continuing basis for the succeeding series of discussions. In order to acquaint the reader both with the detailed workings of certain procedures and the broad applicability of others, examples of construction work ranging from modest to comprehensive in extent are needed. To provide examples of both macro and microwork packages, a large-scale project consisting of several separate segments or subprojects serves as the Example Project.

Two or more of the segments of the Example Project are used for illustrative purposes where a considerable scope of construction activity is needed to present a given management application. An individual subproject is adopted where the level of detail is such that the procedure is best explained by using an example of limited proportions. Several project management actions are presented subsequently using segments of the Example Project as the basis for discussion. Each major management responsibility is the subject of a different chapter. The changes, modifications, revisions, and corrections that are discussed in any one chapter are limited to that chapter and do not carry forward to the next. For purposes of clarity, the methods presented in each chapter are discussed independently of one another and are applied, in turn, to the original, unchanged Example Project.

2.17   THE EXAMPLE PROJECT

The purpose of this section is to describe the Example Project, portions of which will be used throughout this book to explain the several management procedures as they are discussed. The design of the Example Project has been completed, and working drawings and specifications are now available. The owner is a public agency, and the project will be competitively bid by prequalified contractors. Accordingly, the project field management will be carried out by the successful prime contractor. Public notice of the receipt of proposals has been given as required by law, bidding documents are available, and the bidding process is about to begin. The entire Example Project will be awarded by a single contract to the low-bidding prime contractor. Because of the nature of the work involved, unit-price bidding will be used. The contract will contain a time requirement for completion of the construction, and liquidated damages will apply in the event of late contract completion.

The Example Project involves the construction of an earth dam and some appurtenant structures. The completed project will serve a number of purposes, including flood control, irrigation, and recreation. The dam will be constructed across an existing river, the flow of which is highly variable with the season. A permanent reservoir will be formed, the extent of which will vary considerably during the year. The job site covers a considerable geographical area that is undeveloped and unpopulated. The flowchart in Figure 2.1 depicts the overall Example Project and the principal operations that will be covered by the construction contract. This figure shows the major aspects of the work, the general sequence of which will be as follows. One of the first operations must be the diversion of the river away from the dam site area. The borrow areas from which the earth dam material is to be obtained are located some distance from the dam site and must be stripped of surface soil and vegetation. Haul roads between the dam and the borrow areas must be developed, keeping grades to a minimum and with hard, smooth riding surfaces. After the river diversion, borrow development, and haul roads have been accomplished, construction of the dam itself can proceed.

While this preparatory work and the dam are underway, other segments of the Example Project can be progressing simultaneously. A concrete emergency spillway is to be built at a location removed from the main dam itself. The new reservoir area necessitates the relocation of 5 miles of existing natural gas pipeline, and a new bridge must be built where an adjoining highway crosses a reservoir inlet. The closure and removal of the river diversion will be the final major construction operation. The highway bridge and the pipeline relocation will be used to illustrate several construction management applications.

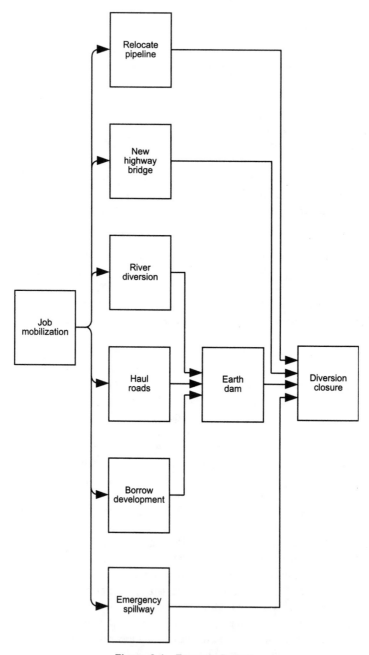

Figure 2.1 Example Project.

3

PROJECT COST ESTIMATING

3.1 THE PROJECT COST SYSTEM

During the design phase of a construction project, the approximate expenses of construction are continually monitored with the objective of keeping the ultimate price of the work within the owner's budget. On completion of design, the field cost-control system is initiated by making a final, detailed cost estimate of the entire work. This estimate is then reduced to a working construction budget.

During the construction process, cost-accounting methods (discussed in Chapter 10) are used to retrieve actual construction expenses from the ongoing project. This information is then used for cost-control purposes on that project and for estimating the expenses of future jobs. Additionally, the cost system provides considerable information pertinent to project financial control, the subject of Chapter 11. This chapter discusses cost-estimating procedures and how the final project budget is obtained.

3.2 PRELIMINARY COST ESTIMATES

Preliminary estimates of future construction expenditures, made during the project planning and design phases, are necessarily approximate because they are compiled while the project is not yet completely defined. Making such conceptual estimates is an art quite different from determining the final detailed estimate of construction costs.

Fundamentally, all conceptual price estimates are based on some system of gross unit costs that have been obtained from previous construction work. These unit costs are extrapolated forward in time to reflect current market conditions, project location, and the peculiar character of the job now under consideration. Some methods commonly used to prepare preliminary estimates are these:

Cost per Function Estimate.

An analysis based on the estimated expenditure per unit of use, such as cost per patient, student, seat, or car space. Construction expense may also be approximated as the average outlay per unit of a plant's manufacturing or production capacity.

Index Number Estimate.

This involves estimating the price of a proposed structure through updating the construction cost of a similar existing facility. This is done by multiplying the original construction cost of the existing structure by a national price index that has been adjusted to local conditions such as weather, labor expense, transportation, and site location. A price index is the ratio of present construction cost to the original construction outlay for the type of structure involved. Many forms of price indexes are available in various trade publications.

Square-foot Cost Estimate.

An approximate cost obtained by using an estimated price for each square foot of gross floor area.

Cubic-foot Cost Estimate.

An estimate based on an approximated expenditure for each cubic foot of the total volume enclosed.

Panel Unit-cost Estimate.

An analysis based on unit costs per square foot of floors, perimeter walls, partition walls, and roof.

Parameter Cost Estimate.

An estimate involving unit costs, called parameter costs, for each of several different building components or systems. The prices of site work, foundations, floors, exterior walls, interior walls, structure, roof, doors, glazed openings, plumbing, heating and ventilating, electrical, and other items are determined separately by the use of estimated parameter costs. These unit expenses can be based on dimensions or quantities of the components themselves or on the common measure of building square footage.

Partial Takeoff Estimate.

An analysis using quantities of major work items taken from partially completed design documents. These are priced using estimated unit prices for each work item taken off.

3.3   THE FINAL COST ESTIMATE

The final cost estimate of a project is prepared when finalized working drawings and specifications are available. This detailed estimate of construction expense is based on a complete and detailed survey of work quantities required to accomplish the work. The process involves the identification, compilation, and analysis of the many items of cost that will enter into the construction process. Such estimating, which is done before the work is actually performed, requires careful and detailed study of the design documents together with an intimate knowledge of the prices, availability, and characteristics of materials, construction equipment, and labor.

It must be recognized that even the final construction estimate is of limited accuracy and that it bears little resemblance to the advance determination of the production costs of mass-produced goods. By virtue of standardized conditions and close plant control, a manufacturer can arrive at the future expense of a unit of production with considerable precision. Construction estimating, by comparison, is a relatively crude process. The absence of any appreciable standardization, together with a myriad of unique site and project conditions, make the advance computation of exact construction expenditures a matter more of accident than design. Nevertheless, a skilled and experienced estimator, using cost-accounting information gleaned from previous construction work of a similar nature, can do a creditable job of predicting construction disbursements despite the project imponderables normally involved. The character or location of a construction project sometimes presents unique problems, but some basic principles for which there is precedent almost always apply.

When pricing a job of some size, there will undoubtedly be more than one person involved with the quantity takeoff and pricing phases. The term "estimator" is used herein to refer to whomever may be involved with the estimating procedure being discussed.

There are probably as many different estimating procedures as there are estimators. In any process involving such a large number of intricate manipulations, variations naturally result. The form of the data generated, the sequential order followed, the nature of the elementary work classifications used, the mode of applying costs — all are subject to considerable diversity. Individual estimators develop and mold procedures to fit their own context and to suit their own preferences. But here, rather than attempt any detailed discussion of estimating methods, only the general aspects of construction estimating are presented.

3.4   THE HIGHWAY BRIDGE

To illustrate the workings of several major aspects of project management, including the cost system, it will be useful to have a construction job large enough to be meaningful, but not so large that its sheer size will obscure the basic objectives. The highway bridge, a segment of the Example Project, has been selected to serve as the basis for discussion of a project estimating and cost system. Although the highway bridge would be estimated and bid as a part of the total Example Project, it has been isolated here for demonstration purposes.

The Highway Bridge. The structure to be erected is a single-span vehicular bridge which will cross a small ravine. The bridge is of a deck-girder type and is of composite steel-concrete construction. Figures 3.1 and 3.2 show the bridge profile and a transverse section. The two abutments are of reinforced concrete, each consisting of a breast wall and two wing walls. Each abutment rests on a heavy concrete footing supported by twenty-eight, 40-foot long H-section steel piles. The 10-inch-thick reinforced concrete paving slab is supported by seven, W36 X 150 steel floor girders. A steel guardrail is required the length of the bridge along each side. All exposed structural concrete is to be given a rubbed finish and certain bridge surfaces are to be painted.

The owner of the Example Project is a public agency and the entire work, including the highway bridge, is to be competitively bid on the basis of unit prices. The final estimate and working job budget that will be obtained in the sections following are those prepared by the prime contractor for its bidding and cost control purposes. The design has been completed and bidding documents, including final-ized drawings and specifications, are in the hands of the bidding contractors.

Figure 3.3, the proposal form to be used for the bidding of the highway bridge, shows 12 bid items and the engineer's quantity estimate of each. The Unit Price, column 5, and Estimated Amount, column 6, are filled in by the bidding contractor after the estimating process has been completed (see Figure 3.9).

3.5 THE QUANTITY SURVEY

The first step in preparing the final price estimate of the highway bridge is the preparation of a quantity survey. This survey is simply a detailed compilation of the nature and quantity of each work type required. Taking off quantities is done in substantial detail, the bridge being divided into many different work types or classifications. Such a takeoff is made where unit-price bidding is involved, even though estimated quantities of each bid item are customarily provided by the architect–engineer with the bidding documents. A basic reason for making a quantity survey for unit-price bidding is that most bid items cannot be priced without breaking the work down into smaller subdivisions. Another reason is that the architect–engineer's quantity estimates, such as those given in column 4 of Figure 3.3, are specifically stated to be approximations only. Making the quantity survey also provides the intimate familiarity with job requirements so necessary for realistic project expensing.

The estimator takes the dimensions and numbers of units of each work type from the design drawings, entering these on quantity sheets and extending these figures into totals. The summarized results of the quantity survey on the highway bridge are shown in Figure 3.4. This figure does not show any quantities for painting. The reason for this is that the prime contractor will normally limit its takeoff to work items it might carry out with its own forces. The contractor intends to subcontract the painting on the highway bridge and thus did not take off quantities of this specialized work category.

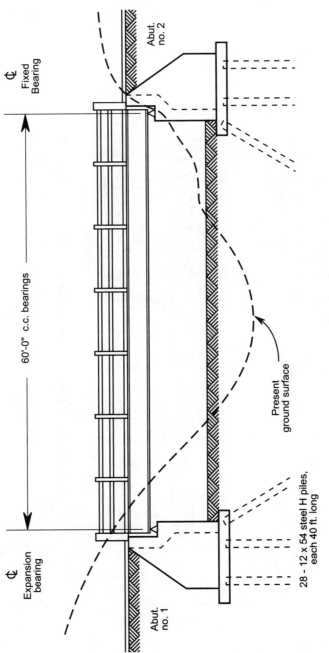

Figure 3.1 Highway bridge, profile.

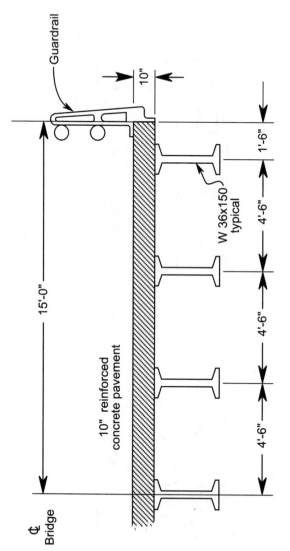

Figure 3.2 Highway bridge, transverse section.

UNIT PRICE SCHEDULE					
Item No. [1]	Description [2]	Unit [3]	Estimated Quantity [4]	Unit Price [5]	Estimated Amount [6]
1	Excavation, unclassified	cy	1,667		
2	Excavation, structural	cy	120		
3	Backfill, compacted	cy	340		
4	Piling, steel	lf	2,240		
5	Concrete, footings	cy	120		
6	Concrete, abutments	cy	280		
7	Concrete, deck slab, 10 in.	sy	200		
8	Steel, reinforcing	lb	90,000		
9	Steel, structural	lb	65,500		
10	Bearing plates	lb	3,200		
11	Guardrail	lf	120		
12	Paint	ls	job		
			Total Estimated Amount =		

Figure 3.3 Highway bridge, bid form.

A comparison of Figures 3.3 and 3.4 will show that the architect–engineer's estimate and the contractor's takeoff are the same insofar as the quantities of the individual bid items are concerned. Quantities estimated by the architect–engineer and by the bidding contractors do not always check and, at times, the differences can be substantial. However, for such a relatively precise work package as the highway bridge, it is reasonable to assume that agreement of the quantity figures would be relatively close in all cases.

3.6 MANAGEMENT INPUT

Early in the estimating process, certainly before the work is priced, a number of important management decisions must be made concerning the project and how construction operations are to be conducted. When the job is being costed, the estimator must exert every effort to price each work type as realistically as possible. To do this, major decisions must be made concerning project organization, the major construction methods to be used, the sequential order of operations, and what construction equipment will be utilized. These four considerations require management attention and are of consuming importance to the bidding process.

A new project cannot be intelligently priced until some major management determinations have been made concerning the conduct of the work. It is clear, therefore, that there must be some regular and usual procedure for the estimator to precipitate such decision making. An effective means of doing this is to have a prebid meeting of knowledgeable company personnel who have the authority to make decisions and binding commitments. If at all possible, the group should include the proposed project manager and the field superintendent. At this meeting, details of the job are discussed, job requirements are reviewed, alternative choices are evaluated, and decisions are made.

Work Quantities			
Cost Code	Work Type	Unit	Quantity
	Sitework:		
02220.10	Excavation, unclassified	cy	1,667
02222.10	Excavation, structural	cy	120
02226.10	Backfill, compacted	cy	340
02350.00	Piledriving rig, mobilization & demobilization	ls	job
02361.10	Piling, steel, driving	lf	2,240
	Concrete		
03150.10	Footing forms, fabricate	sf	360
03150.20	Abutment forms, prefabricate	sf	1,810
03157.10	Footing forms, place	sf	720
03159.10	Footing forms, strip	sf	720
03157.20	Abutment forms, place	sf	3,620
03159.20	Abutment forms, strip	sf	3,620
03157.30	Deck forms, place	sf	1,800
03159.30	Deck forms, strip	sf	1,800
03200.10	Steel, reinforcing, place	lb	90,000
03251.10	Concrete, deck, saw joints	lf	60
03311.10	Concrete, footings, place	cy	120
03311.20	Concrete, abutments, place	cy	280
03311.30	Concrete, deck, place & screed	sy	200
03345.30	Concrete, deck, finish	sf	1,800
03346.20	Concrete, abutments, rub	sf	1,960
03370.20	Concrete, abutments, curing	sf	3,820
03370.30	Concrete, deck, curing	sf	1,800
	Metals		
05120.00	Steel, structural, place	lb	65,500
05520.00	Guardrail	lf	120
05812.00	Bearing plates	lb	3,200

Figure 3.4   Highway bridge, work quantities.

3.7   FIELD SUPERVISION

Before pricing the job, it is always good practice to identify the top supervisors who will be assigned to it. The reason for this is not only to establish specific salary requirements, but also to recognize that most job supervisors do better in terms of construction costs on some portions of a project than on others. Some superintendents are known to be very good at "getting a job out of the ground," but do not perform as well during later phases of the construction. A given supervisor may be experienced with one equipment type, but not with another. The pricing of a project must take into account the special abilities of key field personnel.

The matching of field supervisory talents with the demands of a particular project is one of the most important management actions to be taken insofar as that project is concerned. All the management systems in the world cannot overcome the handicap of poor supervision. The importance of an experienced, skilled, and energetic field supervisory team cannot be overemphasized.

3.8 CONSTRUCTION METHODS

Seldom is there a job operation that can be performed in only one manner. Almost invariably, there is more than one way to accomplish a given item of work. The choice is preferably made after evaluating the time and cost characteristics of the feasible alternatives. This does not suggest that decisions must be made concerning alternative ways of doing every construction operation. As a matter of fact, custom, together with company equipment and experience, automatically make most such choices for the estimator. However, there are sometimes choices to be made concerning construction operations that are important enough to justify the making of detailed comparative studies.

Many examples of decisions concerning construction methods can be cited. Procedures to be followed in underpinning an adjacent structure, how best to brace an excavation, what method of scaffolding to be used, how to dewater the site — all of these involve judgments of major import to the conduct of the work. The proper evaluation of alternatives can require considerable time and extensive engineering studies. It is obvious, however, that the principal construction procedures to be used must be identified before the job can be intelligently priced.

3.9 GENERAL TIME SCHEDULE

When a new project is being estimated, it is necessary that a general plan and operational time schedule be devised. Estimators customarily do this, although often in an informal and almost subconscious way during the takeoff stage. Small jobs may require little investigation in this regard, but larger projects deserve more than a cursory time study. Time is of first-rate importance on all projects. One reason for this is that most contracts impose a required completion date on the contractor.

An approximate construction schedule is also important for project pricing purposes. Many items of job overhead expense are almost directly dependent on the duration of the construction period. When a calendar of work operations is prepared, the time periods required for each of the major job parts can be established as well as the kinds of weather to be anticipated. This provides invaluable information to the estimator concerning equipment and labor productivity, cold weather operations, necessity of multiple shifts or overtime, and other such matters.

Devising a general job plan and time schedule must start with a study of project requirements. This will enable an approximation to be made of the times necessary to accomplish each of the major job segments. The sequence in which these segments must proceed is next established. The result is a bar chart, a series of bars plotted against a horizontal time scale, showing the completion date of the overall project and the approximate calendar times during which the various parts of the job will proceed. Each bar represents the beginning, duration, and completion of some designated segment of the total project. Together, the bars make up a time schedule for the entire job. If the bar chart completion date is not consonant with owner requirements, the estimator will have to rework the job plan.

	June	July	August	September	October
Excavation & backfill	▨▨		▨▨		
Piling		▨			
Abutment No. 1		▨▨			
Abutment No. 2			▨▨		
Steel girders			▨		
Concrete deck				▨▨	
Finishing operations				▨▨	

Figure 3.5 *Highway bridge, general time schedule.*

Although the bar chart has had the advantage of some general planning, it cannot be regarded as the equivalent of a detailed network analysis. However, the development of a CPM job schedule requires considerable time and effort. Consequently, on a competitively bid job, a contractor will not usually make a full-scale time study until it has been proclaimed the successful bidder and awarded the contract. This practical fact emphasizes the need for making up a reasonably accurate general time schedule of the project during the bidding period.

Figure 3.5 shows the general time schedule worked out for the highway bridge. It indicates that a construction period of about 15 weeks will be necessary. Subsequent refinement and development of the job plan will undoubtedly disclose imperfections. Nevertheless, if the development of Figure 3.5 has received adequate consideration, it will serve as an acceptably accurate job picture for pricing purposes. A considerable amount of job planning has already occurred to develop the job schedule to this stage.

3.10 CONSTRUCTION EQUIPMENT

Projects that are of the highway, heavy, or utility category normally require considerable amounts of construction equipment for their accomplishment. A substantial proportion of the total cost of these projects is associated with such equipment. Equipment expenses, however, are highly variable with the type and size of the individual unit. Commensurately, the detailed pricing of equipment on the bridge project cannot proceed very far until equipment selection decisions have been made in fairly specific terms. It is essential that the estimator be able to price the job with reasonable assurance that he is doing so on the basis of the equipment types that will actually be used during construction operations.

To illustrate the workings of equipment decisions, consider the equipment commitments made for the highway bridge. A decision to use transit-mix concrete obtained from a commercial concern obviates the need for a field concrete plant. For pouring concrete, placing structural steel, and driving steel piles, a 50-ton crane equipped with an 80-foot boom will be used. A 7200-foot-pound double-acting hammer and a 900 cubic foot per minute portable air compressor with hose and connections will be used for the pile driving. A low-boy truck trailer and a 25-ton crane will be needed for transport and assembly of the pile driving rig. A crawler tractor with bulldozer blade will do the unclassified excavation and a 1-cubic yard backhoe will be used for the structural excavation. A flat bed truck, troweling machine, and concrete saw will complete the list of larger equipment needs. Smaller equipment such as concrete vibrators and assorted small tools will be provided as needed. These are the kinds of advance equipment decisions that must be made during the estimating period.

3.11 SUMMARY SHEETS

After the quantity survey has been completed and decisions concerning methods and equipment have been made, the total quantities are transferred from the quantity sheets to "summary sheets" for pricing purposes. On lump-sum projects, it is standard practice that all quantities of work types pertaining to a single construction classification, such as concrete, be transferred to, and be priced on a concrete summary sheet. Similar summary sheets are prepared for the other work classifications such as excavation, concrete forms, masonry, and carpentry. On a unit-price job such as the highway bridge, each summary sheet lists the work types necessary to accomplish the total quantity of a single bid item and may include several different classifications of work. Figure 3.6 is the summary sheet for Bid item No. 6, Concrete, abutments, on the highway bridge. The summary sheets for the other 11 bid items are presented in Appendix A. In Figure 3.6 and in Appendix A, each work type is designated by an identifying number. These are the contractor's standard cost-account numbers which are basic to the workings of its cost-accounting system. When the takeoff is being compiled for a new project, the work is broken down into the standard elementary classifications established by the system. Each of these standard work types bears a unique cost-account number. The coding system will be discussed further in Chapter 10.

3.12 MATERIAL COSTS

It is customary for the contractor to solicit and receive specific price quotations for most of the materials required by the job being priced. Exceptions to this generality would be stock items such as plyform, nails, and lumber that the contractor purchases in large quantities and of which a running inventory is maintained. Written quotations for special job materials are desirable so that such important considerations as prices, freight charges, taxes, delivery schedule, and guarantees are explicitly understood. Most material suppliers tender their quotations on printed forms that include stipulations pertaining to terms of payment and other considerations. On the highway bridge, the contractor will receive during the bidding period written price quotations from material dealers covering specific job materials such as transit-mix concrete, structural and reinforcing steel, steel pilings, and guardrail. Consequently, if the quantity survey has been done with some exactness, materials can usually be priced quite accurately.

Material costs, when entered on the summary sheets, must all be on a common basis: for example, delivered to the job site and without sales tax. Prices as entered will ordinarily include freight, drayage, storage, and inspection. It is common practice to enter material prices without tax, adding this as a lump-sum amount on the final recap sheet (see Section 3.25).

It is not unusual that the owner provides certain materials to the contractor for use on the project, although this is not the case on the highway bridge. In such cases, contractors need not add this material price into their estimates. However, all other charges associated with the material such as handling and installation expense must be included.

3.13 LABOR COSTS

The real challenge in pricing construction work is the determination of labor and equipment expenditures. These are the categories of construction expense that are inherently variable and the most difficult to estimate accurately. To do an acceptable job of establishing these outlays, the estimator must make a complete and thorough job analysis, maintain a comprehensive library of costs and production rates from past projects, and obtain advance decisions about how construction operations will be conducted.

Contractors differ widely in how they estimate labor costs. Some choose to include all elements of labor expense into a single hourly rate. Others evaluate direct labor cost separately from indirect cost. Some contractors compute regular and overtime labor costs separately, while others combine scheduled overtime with straight time into an average hourly rate. Some evaluate labor charges using production rates; others use labor unit costs. There are usually good reasons why a given contractor evaluates its labor expense as it does, and there certainly is no single correct method that must be followed. The procedures described in this chapter are commonly used and are reasonably representative of general practice.

Basic to the determination of the labor cost associated with any work category is the production rate. An example of a production rate is illustrated by Figure 3.6 in which the labor expense of pouring the abutment concrete on the bridge project is determined. Reference to Figure 3.6 shows that a placement rate of 10 cubic yards of concrete per hour is used as the production rate for the prescribed concrete crew. The direct labor cost to pour the entire 280 cubic yards of abutment concrete is obtained by multiplying the time required to pour the concrete (32 hours) by the direct hourly wage rate of the entire crew ($145.43) giving $4,654. To this must be added indirect labor costs of $1,722 giving a total labor charge of $6,376 to accomplish this particular item of work. The distinction between direct and indirect labor expenses is discussed in Section 3.14.

The most reliable source of labor productivity information is obtained from cost-accounting reports compiled from completed projects. Labor outlay information is also available from a wide variety of published sources. However, while information of this type can be very useful at times, it must be emphasized that labor productivity differs from one geographical location to another and is variable with season and many other job factors. Properly maintained labor records from recent jobs completed in the locale of the project being estimated reflect, to the maximum extent possible, the effect of local and seasonal conditions.

3.14 INDIRECT LABOR COSTS

Direct labor cost is determined from the workmen's basic wage rates; that is, the hourly rates used for payroll purposes. Indirect labor costs are those expenses which are additions to the basic hourly rates and which are paid by the employer. Indirect labor expense involves various forms of payroll taxes, insurance, and employee fringe benefits of wide variety. Employer contribution to social security, unemployment insurance, workers' compensation insurance, and contractor's public liability and property damage insurance are all based on payrolls. Employers in the construction industry typically provide for various kinds of fringe benefits such as pension plans, health and welfare funds, employee insurance, paid vacations, and apprenticeship programs. The charge for these benefits is customarily based on direct payroll costs. Premiums for workers' compensation insurance and most fringe benefits differ considerably from one craft to another.

Indirect labor costs are substantial in amount, often constituting a 35 to 55 percent addition to direct payroll expenses. Exactly when and how indirect labor costs are added into a project estimate are unimportant so long as it is done. For estimating purposes, total labor outlay can be computed in one operation by using hourly labor rates which include both direct and indirect costs. However, this procedure may not interrelate well with labor cost-accounting methods. For this reason, direct and indirect labor charges are often computed separately when job prices are being estimated. One commonly used scheme is to add a percentage allowance for indirect cost to the total direct labor expense, either for the entire project or for each major work category. Because of the appreciable variation of indirect costs from one classification of labor to another, it may be preferable to compute indirect labor cost at the same time that direct labor expense is obtained for a given work type. This is the method followed in Figure 3.6 and Appendix A.

3.15 LABOR UNIT COSTS

Direct labor cost was computed in Section 3.13 for a given work type using hourly crew payroll expense together with the work quantity and appropriate production rate. There is a widely used alternative to this procedure that involves the use of "labor unit costs." A labor unit cost is the direct labor expense per unit of production of a work type. To illustrate, reference is again made to concrete placing in Figure 3.6. This figure shows that 32 hours of crew time are required to place 280 cubic yards of abutment concrete at a total direct labor outlay of $4,654. Dividing $4,654 by 280 yields $16.62 This value of $16.62 is a labor unit cost; it is the average direct labor

SUMMARY SHEET

Job: Highway Bridge
Bid Item No. 6: Concrete, abutments
Estimator: GAS

Cost Code	Work Type	Quantity	Unit	Calculations	Labor Cost Direct	Labor Cost Indirect	Equipment Cost	Material Cost	Total Cost
03150.20	Abutment forms, prefabricate	1,810	sf	**Labor:** Labor unit cost = $1.10/sf 1,810 x $1.10 = $1,991 **Material:** Plyform: 10% waste, 2 uses, 50% salvage 1,810 x 1.1 x$0.38 x 0.5 =$378 Lumber: 4 fbm/sf, 2 uses, 50% salvage 1,810 x 4 x $0.35 x .5 = $1267	$1,991	$916		$378 $1,267	$2,907 $378 $1,267
				Total this account	$1,991	$916	$0	$1,645	$4,552
03157.20	Abutment forms, place & strip	3,620	sf	**Labor:** Labor unit cost = $2.31/sf 3,620($2.31) = $8,362 (Approx. 70% for placing, 30% for stripping) **Equipment:** 50 ton crane for 3 days 3 x 8 x $108 = $2,592 (Approx. 70% for placing, 30% for stripping) **Material:** Nails, form ties, coating 3,620($0.10) = $362	$8,362	$3,696	$2,592	$362	$12,058 $2,592 $362
				Total this account	$8,362	$3,696	$2,592	$362	$15,012
03311.20	Concrete, abutment place	280	cy	**Labor:** 1 foreman @ $17.20 = $17.20 1 mason @ $14.40 = $14.40					

Account	Description	Qty	Unit	Calculation					
				6 laborers @ $11.60 = $ 69.60					
				1 operator @ $15.95 = $15.95					
				1 oiler @ $12.08 = $12.08					
				1 carpenter @ $16.20 = $16.20					
				Crew hourly rate = $145.43					
				Production rate = 10 cy/hr					
				280 ÷ 10 = 28 hrs, say four 8 hr days					
				32 x $145.43 = $4,654	$4,654	$1,722			$6,376
				Equipment:					
				50 ton crane @ $108.00					
				2 vibrators & buckets @ $14.50					
				Equipment hourly rate = $122.50					
				280 ÷10 x $122.50 = $3,430			$3,430		$3,430
				Material:					
				Transit mix @ $49.50/cy, 5% waste					
				280 x 1.05 x $49.50 = $14,553				$14,553	$14,553
03346.20	Concrete, abutment rub	1,960	sf	Total this account	$4,654	$1,722	$3,430	$14,553	$24,359
				Labor:					
				Production rate = 16.67 sf/mh					
				1,960 ÷ 16.67 x $14.40 = $1,693	$1,693	$762			$2,455
				Material:					
				Material unit cost = $0.06/sf					
				1,960 x $0.06 = $118				$118	$118
				Total this account	$1,693	$762	$0	$118	$2,573
				Labor:					
				Production rate = 500 sf/hr					
				3,820 ÷ 500 x $14.40 = $110	$110	$42			$152
				Material:					
				Curing compound covers 270 sf/gal					
				3,820 ÷ 270 = 15 gal.					
				15 x $4.60 = $69				$69	$69
03370.20	Concrete, abutment curing	3,820	sf	Total this account	$110	$42	$0	$69	$221
Total Bid Item No. 6					$16,810	$7,138	$6,022	$16,747	$46,717
						$23,948			

Figure 3.6   Highway bridge, bid-item summary sheet.

charge of pouring 1 cubic yard of abutment concrete. Thus, the direct labor cost of pouring the abutment concrete could have been computed in Figure 3.6 simply by multiplying the total of 280 cubic yards of abutment concrete by the labor unit expense of $16.62 and the same value of $4,654 would have been obtained.

The use of labor unit costs in estimating practice is usually limited to the determination of direct labor expense. Once the direct labor cost is computed for a given work category, the applicable indirect labor cost is then determined by multiplying the direct labor expense by the appropriate percentage figure. Labor unit prices are used in Figure 3.6 to compute the direct labor costs of fabricating the abutment forms and for placing and stripping the abutment forms.

When labor unit costs are being used, care must be exercised that they are based on the appropriate levels of work productivity and the proper wage rates. Also, estimators must be very circumspect when using labor unit costs that they have not developed themselves. For the same work items, different estimators will include different expenses in their labor unit costs. It is never advisable to use a labor unit cost derived from another source without knowing exactly the categories of expense it does and does not include.

3.16   EQUIPMENT COST ESTIMATING

Unfortunately, the term "equipment" does not have a unique connotation in the construction industry. A common usage of the word refers to scaffolding, hoists, power shovels, paving machines, and other such items used by contractors to accomplish the work. However, equipment is also used with reference to various kinds of mechanical and electrical furnishings that become a part of the finished project such as boilers, escalators, electric motors, and hospital sterilizers. In this text, equipment refers only to the contractor's construction equipage. The term "materials" will be construed to include all items that become a part of the finished structure, including electrical and mechanical plant.

Equipment costs, like those of labor, are difficult to evaluate and price with any precision. Equipment accounts for a substantial proportion of the total construction expense of most engineering projects, but is less significant for buildings. When the nature of the work requires major items of equipment such as earth-moving machines, concrete plants, and truck cranes, detailed studies of the associated costs must be made. Expenses associated with minor equipment items such as power tools, concrete vibrators, and concrete buggies are not normally subjected to detailed study. A standard expense allowance for each such item required is included, usually based on the time that it will be required on the job. The cost of small tools, wheelbarrows, water hose, extension cords, and the like are covered by a lump-sum allowance sometimes obtained as a small percentage of the total labor cost of the project.

To estimate the expense of major equipment items as realistically as possible, early management decisions must be made concerning the equipment sizes and types required and the manner in which the necessary units will be provided to the project. A scheme sometimes used when the duration of the construction period will be about equal to the service life of the equipment is to purchase all new or renovated equipment for the project and sell it at the cessation of construction activities. The difference between the purchase price and the estimated salvage value is entered into the job estimate as a lump-sum equipment expenditure.

Equipment is frequently rented or leased. Rental can be especially advantageous when the job site is far removed geographically from the contractor's other operations, for satisfying temporary peak demand, or for providing specialized or seldom-used equipment. Leasing is a common and widely used means of acquiring construction equipment and may be a desirable alternative to equipment ownership. Leasing can improve a contractor's working capital position by avoiding having its funds tied up in fixed assets. Under certain circumstances, lease payments compare favorably with ownership expense. Many leases provide that at the expiration of the lease period the contractor has a purchase option if it has continuing use for the machine and if it is worth the additional payment. Lease agreements for construction equipment normally extend for periods of one year or more, whereas renting is usually of shorter term. Charges associated with the rental or lease of equipment items are figured into the job by applying the lease or rental rates to the time periods that the equipment will be needed on the project.

Where purchase and sale, rental, or lease is involved, equipment operating expenses must also be computed and included in the project estimate. Operating costs include charges such as fuel, oil, grease, filters, repairs and parts, tire replacement and repairs, and maintenance labor and supplies. There is some difference of opinion about whether the wages of equipment operators should be included in the equipment operating cost. Some contractors prefer to regard the labor associated with equipment operation as a labor rather than an equipment cost. Others include the labor as a part of equipment operating expense. Logically, it would seem preferable to treat equipment-operating labor as any other labor cost rather than include it with equipment-operating expense. For purposes of discussion herein, equipment operators' wages are treated as a labor cost and are not included as a part of equipment expense.

The most common procedure is for contractors to own their own equipment. Equipment charges, under these circumstances, are customarily expressed as the sum of ownership expense and operating costs. Ownership expenses are those of a fixed nature and include depreciation, interest on investment or financing charges, taxes, insurance, and storage. Operating expenses have been defined previously.

3.17 EQUIPMENT EXPENSE

As described in Section 3.13, labor direct costs are computed from work quantities by combining a labor production rate with the applicable hourly wage scales. Most equipment costs are calculated in much the same fashion except, of course, equipment production rates and equipment hourly costs must be used. The hourly wage rates of various labor categories are immediately determinable, usually from applicable labor contracts, prescribed prevailing wage rates, or established area practice. This is not true for equipment. Contractors must establish their own equipment hourly rates as well as their equipment production rates. For most items of operating equipment, ownership, lease, or rental expense is combined with operating costs into an estimated total charge per operating hour. Power shovels, tractor scrapers, and ditchers are examples of equipment whose expenses are usually expressed in terms of hourly rates.

There are some classes of construction equipment, however, for which it is more appropriate to express costs in terms of time units other than operating hours. The charge for commercial prefabricated concrete forms might be better spread over an estimated number of reusages. Items such as towers and scaffolding are required at the job site on a continuous basis during particular phases of the work, and operating hours have no significance in such cases. Costs in terms of other time units such as calendar months are more appropriate for such equipment items. The charges for some classes of production equipment are frequently expressed in terms of expense per unit of material produced. Portland cement concrete-mixing plants, asphalt-paving plants, and aggregate plants are familiar instances of this.

Move-in, erection, dismantling, and move-out expenses, also called mobilization and demobilization costs, are entirely independent of equipment-operating time and production and are not, therefore, included in equipment hourly rates. These equipment expenses are separately computed for inclusion in the estimate. Mobilization of the pile driving rig for the highway bridge is an example of this type of equipment cost and is shown in Bid Item No. 4 of Appendix A.

3.18   DETERMINATION OF EQUIPMENT EXPENSE RATES

The determination of the expense rates for the equipment items to be used on a project being estimated is an important matter and requires considerable time and effort. It must be remembered that estimators are always working in the future tense and that the equipment rates that they use in their estimates are their best approximations of what such costs will turn out to be when the project is actually being built. Equipment expense rates are approximations at best and must be regarded as such.

The standard way in which future equipment expense is determined is through the use of historical equipment-cost records as contained in a contractor's equipment-accounting system. The source of ownership and operating cost data for a specific piece of equipment is its ledger account. An important part of a contractor's general accounting system is the equipment accounts where actual ownership and operating expenses are recorded as they are incurred. A separate account is often set up for each major piece of equipment. This account serves to maintain a detailed and cumulative record of the use of, and all expenses chargeable to, that equipment item. All expenditures associated with that piece of equipment, regardless of their nature or the project involved, are charged to that account. These expenses include depreciation, investment costs, operating outlays, repairs, parts, overhauls, and painting. Also maintained is a cumulative record of that equipment item's usage in the field. These data constitute a basic resource for reducing ownership and operating expense to a total equipment cost rate. The sum of ownership and operating expense expressed as a cost per unit of time (operating hour, week, month) is often referred to either as the "budget rate" or the "internal rental rate" of an equipment item. This latter term refers to the usual construction accounting practice where equipment time on the project during construction is charged against the job at that rate.

The preceding discussion assumes that equipment accounting is done on an individual machine basis. However, contractors vary somewhat in how they maintain their equipment accounts, some preferring to keep equipment costs by categories of equipment rather than by individual unit. These firms use a single account for all equipment items of a given size and type and compute an average budget rate based on the composite experience with all of the units included. Thus the same expense rate is applied for any unit of a given equipment type regardless of differences in age or condition. Because the actual costs and productivity of individual units can vary substantially, working with individual equipment items would seem to have merit and is the basis for the discussions herein.

There are many external sources of information concerning ownership and operating costs of a wide range of construction equipment. Manufacturers, equipment dealers, and a large assortment of publications offer such data. It must be realized, however, that these are typical or average figures and that they must be adjusted to reflect contractor experience and methods and to accommodate the specific circumstances of the project being estimated. Climate, altitude, weather, job location and conditions, operator skill, field supervision, and other factors can and do have a profound influence on equipment expense. When a new piece of equipment is procured for which there is no cost history, equipment expense must be estimated using available sources of information considered to be reliable.

In the case of the highway bridge, reference to Figure 3.6 shows the budget rate for the 50 ton crane to be $108 per operating hour. This does not include the wages of the operator, this expense being included in labor cost.

3.19 EQUIPMENT PRODUCTION RATES

In addition to equipment cost, equipment production rates are also often needed for the computation of equipment expense in a construction estimate. Paralleling the case of labor, applying equipment hourly expenses and production rates to job quantities enables the estimator to compute total equipment charges for the project. "Equipment unit costs" can also be determined, which are equipment costs per unit of production. Equipment production rates, like those of labor, are subject to considerable variation and are influenced by a host of job site conditions. In addition, some equipment production rates must be computed using specific job conditions such as haul distances, grades, and rolling resistance. Estimators must consider and evaluate these factors when they are pricing a new project.

There are several sources of equipment production information. Probably the most reliable are cost-accounting records from past projects. Advice from the equipment operators themselves can be very useful at times. If a new piece of equipment is involved with which there has been no prior experience, production information provided by the equipment manufacturer or dealer can be of assistance. There are many rules of thumb and published sources of information concerning average equipment production rates. Stopwatch "spot checks" made to obtain the productivity of specific equipment items on past projects can be of value. In this regard, it should be mentioned that production rates of labor and equipment used for estimating purposes should be average figures taken over a period of time. Daily job production tends to be variable, and this is a hazard with using spot checks. Job cost-accounting records produce good time-average values while a spot check may unknowingly catch production at a high or low point.

Several equipment production rates are needed for pricing the highway bridge. One such instance is the time rate at which the steel pilings can be driven. Reference to Bid Item No. 4 of Appendix A shows that past company experience indicates a probable driving rate of 70 lineal feet per hour. The hourly equipment expense for the crane, air compressor, pile hammer, and leads totals $177.38 per hour. Using a total project pile footage of 2240 lineal feet, the production rate and hourly equipment expense is used to compute a total equipment cost of $5,676. Equipment unit costs are computed and used just like those of labor. Using pile driving as an example, dividing the hourly equipment rate of $177.38 by the production rate of 70 lineal feet per hour yields an equipment unit cost of $2.53 which is the equipment charge per lineal foot of pile driven.

3.20   BIDS FROM SUBCONTRACTORS

If the prime contractor intends to subcontract portions of the project to specialty contractors, the compilation and analysis of subcontractor bids is an important aspect of making up the final project estimate. Bids from subcontractors sometimes contain qualifications or stipulate that the general contractor is to be responsible for providing the subcontractor with certain job-site services such as hoisting facilities, electricity and water, storage facilities for materials, and many others. Before estimators can identify the low subbid for any particular item of work, they must analyze each bid received to determine exactly what each such proposal includes and does not include. The checking of subbids can be a considerable chore when substantial portions of the project are to be sublet.

On the highway bridge, the general contractor has made an advance decision that the painting will be subcontracted. Painting is a specialty work area which the general contractor is not equipped to do and with which it has had no past field experience. When such a decision is made, the contractor does not compile the cost of doing the work with its own forces. Rather, the lowest subcontract bid received from a responsible subcontractor will be included with the contractor's other expenses. The lowest acceptable painting subcontract bid received was $5,820 and will cover all such services as required by the project drawings and specifications. This is shown by Bid Item No. 12 of Appendix A.

The advance decision to subcontract the painting does not necessarily mean that the general contractor will perform all of the other work with its own forces. Other specialty areas of the bridge may also be subcontracted, this depending upon a number of circumstances. In this regard, the contractor may specifically request subbids from selected subcontractors or it may merely await receipt of such bids that subcontractors voluntarily submit. In any event, the contractor must compile its own cost of doing the work involved and will normally be interested only in those subbids whose amounts are less than its own estimated direct cost.

When the general contractor receives a subbid whose amount is less than its own estimated direct outlay for doing the same work, it cannot accept such a subbid until consideration is given to several factors. Has the contractor had past experience with that subcontractor and can it be expected to carry out its work properly? Does the subcontractor have a history of reliability and financial stability? Is the subcontractor experienced and equipped to do the type of work involved? The general contractor must remember that it is completely responsible by contract with the owner for all subcontracted work as well as that performed by its own forces.

In compiling its bid for the bridge, the prime contractor received a subbid for Bid Item No. 8: Steel, reinforcing. Subcontract bids for reinforcing steel often include only the cost of labor. However, in this case the subbid includes the prices of all materials and labor. The general contractor estimated its own direct expense of providing and placing reinforcing steel for the bridge to be $42,705 as shown in Bid Item No. 8a of Appendix A. The subbid received for the same items was for $40,275, low enough to merit serious consideration. The general contractor has worked with this subcontractor before and found the company to be honest, reliable, and of good reputation. It is a complete bid and does not require the general contractor to provide the subcontractor with specific job-site services. The use of this subbid by the general contractor rather than its own estimated direct cost will reduce the contractor's bid by a significant amount. Consequently, the prime contractor decides to use the reinforcing steel subbid in the final compilation of its proposal to the owner as illustrated by Bid Item 8b of Appendix A.

At this point, it should be pointed out that the bid-item summary sheets have now been completed. From the price information contained in these summary sheets, the contractor will subsequently prepare its working project budget (see Section 3.26) if it becomes the successful bidder. At the moment, however, the estimating process must continue in order to obtain the necessary 12-bid unit prices.

3.21 PROJECT OVERHEAD

Overhead or indirect expenses are outlays which are incurred in achieving project completion, but which do not apply directly to any specific work item. There are actually two kinds of overhead that pertain to a contractor's operations. One of these is project overhead; the other, office overhead which is discussed in the next section. Project overhead, also referred to as job overhead or field overhead, refers to indirect field expenses that are chargeable directly to the project. Some contractors figure their job overhead outlay as a percentage of the total direct job cost, common values for the job overhead allowance being from 5 to 15 percent.

The use of percentages when computing field overhead is not generally considered to be good estimating practice because different projects can and do have widely varying job overhead requirements. The only reliable way to arrive at an accurate estimate of project overhead is to make a detailed analysis of the particular demands of that project. It is standard estimating procedure to list and price each item of indirect expense on a separate overhead sheet. Figure 3.7 is the project overhead

PROJECT OVERHEAD ESTIMATE

Job: Highway Bridge

Estimator: GAS

Overhead Item	Calculations	Amount	Totals
Project manager	$3,300/mo. x 3.5 months x .5 time =	$5,775.00	
Project superintendent	$3,000/mo x 3.5 months =	$10,500.00	
Utilities			
Electricity	$100 /mo.		
Telephone	$220 /mo.		
Fax	$60 /mo.		
	$380 /mo. x 3.5 months =	$1,330.00	
Utility installation charges	(job)	$840.00	
Facilities			
Job office	$200 /mo.		
2 ea. tool sheds	$400 /mo.		
Toilet	$90 /mo.		
	$690 /mo. x 3.5 months =	$2,415.00	
Travel expense	$147/wk x 10 weeks =	$1,470.00	
Water tank & water service	$40/wk x 10 weeks =	$400.00	
Soil & concrete testing	$375/mo x 3.5 months =	$1,312.50	
Scaffolding	$320/mo. x 1 month =	$320.00	
Trash removal	$140/mo. x 3.5 months =	$490.00	
Tire repair	$50/mo. x 3.5 months =	$175.00	
Photographs	$75/mo. x 3.5 months =	$262.50	
Computer	$110/mo. x 3.5 months =	$385.00	
	Subtotal of time variable overhead expenses =		$25,675.00
Surveys	(job)	$600.00	
Project insurance	(job)	$685.00	
First aid	(job)	$100.00	
Sign	(job)	$300.00	
Reproductions	(job)	$300.00	
Fence	$1.25/ft. x 680 feet =	$850.00	
Move In	See calculations, Appendix A	$8,773.00	
Clean Up	See calculations, Appendix A	$3,940.00	
	Subtotal of time constant overhead expenses =		$15,548.00
	Total project overhead =		$41,223.00

Figure 3.7 *Highway bridge, overhead estimate.*

sheet for the bridge project cost estimate. The job overhead amount of $41,223.00 was compiled using an estimated project duration of 15 weeks (from Section 3.9).

3.22   OFFICE OVERHEAD

Office overhead includes general company expenses such as office rent, office insurance, heat, electricity, office supplies, furniture, telephone, legal costs, donations, advertising, office travel, association dues, and office salaries. The total of this overhead expense will usually range from 2 to 8 percent of a contractor's annual business volume. An allowance for such indirect expense must be included in the cost estimate of each new project.

Office overhead is made up of charges which are incurred in support of the overall company construction program and which cannot be charged to any specific project. For this reason, office overhead is normally included in the job estimate as a percentage of the total estimated project expense. The allowance for office overhead can be added as a separate line item in the cost estimate, or a suitable "markup" percentage can be applied or fee established that will include office overhead as well as profit.

3.23 MARKUP

On competitively bid projects, markup or margin is added at the close of the estimating process and is an allowance for profit, plus possibly other items such as office overhead and contingency. Regarding contingency as a separate component of markup is a matter of management philosophy. The profit included in a job bid represents the minimum acceptable return on the contractor's investment. Return on investment is a function of risk, and greater risk calls for a greater profit allowance in the proposal. Whether recognition of risk is in the form of a higher profit percentage or the inclusion of a contingency allowance seems to be a matter of personal preference.

Markup, which may vary from 5 to more than 20 percent of the estimated project cost, represents the contractor's considered appraisal of a whole series of imponderables that may influence its chances of being the low bidder and its making a reasonable profit if it is. Many factors must be considered in deciding a markup figure and each can have an influence on the figure chosen. The size of the project and its complexity, its location, provisions of the contract documents, the contractor's evaluation of the risks and difficulties inherent in the work, the competition, the contractor's desire for the work, the identity of the owner and/or the architect–engineer, and other intangibles can have a bearing on how a contractor marks up a particular job.

The contractor is required to bid under a form of contract that has been specifically written to protect the owner. In an attempt to afford the owner, and often the architect–engineer, protection against liabilities and claims that may arise from the construction process, the drafters of contract documents often incorporate a great deal of "boiler plate," including disclaimers of one sort or another. The writers of contract documents sometimes force the contractor to assume liability for every conceivable contingency, some of which are not subject to its control and which are not rightfully its responsibility. For example, provisions are to be found in some contracts that absolve the owner from all claims for damages arising from project delay, even though caused by its own fault or negligence. Contractors are frequently made to assume full responsibility for any and all unknown physical conditions, including subterranean, which may be found at the construction site. It suffices to say that the markup figure selected must take into account the risks created by such contract provisions.

3.24 CONTRACT BONDS

Many construction contracts, especially those involving public owners, require that the prime contractor provide the owner with a specified form of financial protection against contractor default. Two forms of surety bonds, called contract bonds, are used for this purpose. A contract bond is an agreement, the terms of which provide that a surety company will carry out the contractor's obligations to the.owner if the contractor itself fails to do so. A surety is a party that assumes the legal liability for the debt, default, or failure in duty of another.

By the terms of its contract with the owner, the contractor accepts two principal responsibilities: to perform the objective of the contract and to pay all expenses associated with the work. Where contract bonds are called for, the general contractor is required to provide the owner with a "performance bond" and a "labor and material payment bond." By the terms of these two bonds, the surety guarantees the owner that the work will be completed in accordance with the contract and that all construction costs will be paid should the contractor not perform as promised.

When the bidding documents provide that the successful contractor shall furnish the owner with performance and payment bonds, as is the case with the Example Project, the contractor must purchase these bonds if it is awarded the contract. These bonds are obtained by the contractor from the surety company with which it customarily does business. Sureties are large corporate firms that specialize in the furnishing of many forms of surety bonds, including contract bonds for contractors. The premium charge for these bonds is substantial and varies with the type of work involved and the contract amount. This cost is paid by the contractor and must be included in the price estimate of the project. Because the bond outlay is based on the total contract amount, this is normally the last item of expense to be added into the project estimates. This may be seen in Figure 3.8.

3.25 RECAP SHEET

To calculate the needed bid unit prices, all costs associated with the highway bridge are now brought together in summary form on a recapitulation or recap sheet. Figure 3.8 is the recap sheet for the highway bridge. The expenses of labor, equipment, material, and subcontracts have been entered on the recap sheet from the summary sheets in Figure 3.6 and Appendix A.

On the recap sheet, the direct cost of the entire quantity of each bid item is obtained as the sum of its labor, equipment, material and subcontract expenses. The sum of all such bid-item direct costs gives the estimated total direct cost of the entire project ($216,814.00). To this are added the job overhead, small tools, tax, markup, and the premium charge for the performance and payment surety bonds, giving the total price ($312,724.24). In Figure 3.8, the 15 percent markup includes an allowance for office overhead. Dividing the total project bid by the total direct project cost gives a factor of 1.4424. By multiplying the total direct cost of each bid item by this factor, the total amount of that bid item is obtained. Dividing the total bid cost of each bid item by its quantity gives the bid unit price. The bid unit prices are customarily rounded off to even figures.

The unit prices just computed have been obtained on the basis that each bid item includes its own direct cost plus its pro rata share of the project overhead, small tools, tax, markup, and bond. If these unit prices are now entered without change onto the bid form, this is called "balanced" bidding. For several reasons, a contractor may

RECAP SHEET
Bid Date: April 25, 19—

Job: Highway Bridge   Estimator: GAS

Item No.	Bid Item	Unit	Estimated Quantity	Labor Cost	Equipment Cost	Material Cost	Subcontract Cost	Direct Cost	Bid Total	Bid Unit
1	Excavation, unclassified	cy	1,667	$1,805.00	$619.00	$0.00	$0.00	$2,424.00	$3,496.29	$2.10
2	Excavation, structural	cy	120	$2,067.00	$390.00	$0.00	$0.00	$2,457.00	$3,543.88	$29.53
3	Backfill, compacted	cy	340	$2,219.00	$469.00	$0.00	$0.00	$2,688.00	$3,877.07	$11.40
4	Piling, steel	lf	2,240	$10,956.00	$11,184.00	$30,690.00	$0.00	$52,830.00	$76,199.98	$34.02
5	Concrete, footings	cy	120	$2,313.00	$580.00	$6,560.00	$0.00	$9,453.00	$13,634.65	$113.62
6	Concrete, abutments	cy	280	$23,948.00	$6,022.00	$16,747.00	$0.00	$46,717.00	$67,382.82	$240.65
7	Concrete, deck slab, 10 in	sy	200	$8,086.00	$1,043.00	$4,603.00	$0.00	$13,732.00	$19,806.51	$99.03
8	Steel, reinforcing	lb	90,000	$0.00	$0.00	$0.00	$40,275.00	$40,275.00	$58,091.12	$0.65
9	Steel, structural	lb	65,500	$2,474.00	$1,728.00	$27,052.00	$0.00	$31,254.00	$45,079.58	$0.69
10	Bearing plates	lb	3,200	$1,105.00	$432.00	$2,140.00	$0.00	$3,677.00	$5,303.56	$1.66
11	Guardrail	lf	120	$1,207.00	$432.00	$3,848.00	$0.00	$5,487.00	$7,914.24	$65.95
12	Paint	ls	job	$0.00	$0.00	$0.00	$5,820.00	$5,820.00	$8,394.55	$8,394.55
	Totals			$56,180.00	$22,899.00	$91,640.00	$46,095.00	$216,814.00	$312,724.24	

		Direct Cost
		$216,814.00
Job overhead		$41,223.00
		$258,037.00
Small tools (5% of labor)		$2,809.00
		$260,846.00
Tax	3%	$7,825.38
		$268,671.38
Markup	15%	$40,300.71
		$308,972.09
Bonds		$3,752.16
Total Project Bid		$312,724.24

$$\text{Factor} = \frac{\$312,724}{\$216,814} = 1.4424$$

Figure 3.8 Highway bridge, recap sheet.

UNIT PRICE SCHEDULE					
Item No. [1]	Description [2]	Unit [3]	Estimated Quantity [4]	Unit Price [5]	Estimated Amount [6]
1	Excavation, unclassified	cy	1,667	$2.10	$3,500.70
2	Excavation, structural	cy	120	$29.53	$3,543.60
3	Backfill, compacted	cy	340	$11.40	$3,876.00
4	Piling, steel	lf	2,240	$34.02	$76,204.80
5	Concrete, footings	cy	120	$113.62	$13,634.40
6	Concrete, abutments	cy	280	$240.65	$67,382.00
7	Concrete, deck slab, 10 in.	sy	200	$99.03	$19,806.00
8	Steel, reinforcing	lb	90,000	$0.65	$58,500.00
9	Steel, structural	lb	65,500	$0.69	$45,195.00
10	Bearing plates	lb	3,200	$1.66	$5,312.00
11	Guardrail	lf	120	$65.95	$7,914.00
12	Paint	ls	job	$8,394.55	$8,394.55
				Total Estimated Amount =	$313,263.05

Figure 3.9   *Highway bridge, completed bid form.*

raise the prices on certain bid items and decrease the prices on others proportionately so that the bid amount for the total job remains unaffected. This is called an unbalanced bid. It is assumed here that a balanced bid will be submitted by the contractor. Commensurately, the unit prices determined in Figure 3.8 are now entered in column 5 of Figure 3.9, the unit-price schedule of the bid form. The total estimated amounts in column 6 of Figure 3.9 are obtained by multiplying the unit prices just entered (column 5), by the estimated quantities (column 4). Because of the rounding off of unit prices, the project bid total is slightly different on the unit-price schedule of the bid form ($313,263.05), Figure 3.9, than it is on the recap sheet ($312,724.24), Figure 3.8.

When unit-price proposals are submitted to the owner, the low bidder is determined on the basis of the total estimated amount. Consequently, for contract award purposes, the amount of $313,263.05 in Figure 3.9 is treated just as a lump-sum bid. In cases of error in multiplication or addition by the contractor in obtaining the estimated amounts in column 6 of Figure 3.9, it is usual for the unit prices to control and the corrected total sum to govern.

3.26   THE PROJECT BUDGET

Assuming that our contractor is the successful bidder, it must now restructure its estimate into a more suitable format for subsequent cost control of the actual construction work. This involves the preparation of the "control budget" or the "project budget." This is the detailed schedule of expenses that the project manager will use for cost control purposes during the construction phase. Figure 3.10 is the control budget for the highway bridge project. The work quantities and prices contained in this figure have been extracted from the bid item summary sheets previously discussed and presented. When project cost accounting is discussed in Chapter 10, the actual construction expenses of the highway bridge are compared with the programmed costs contained in the project budget, Figure 3.10. As is discussed in Chapter 10, unit prices are especially useful for making quick and meaningful comparisons of actual and budgeted expense for both labor and equipment. Unit costs of materials are not especially significant except for estimating purposes and are not shown in Figure 3.10.

PROJECT BUDGET

Job: Highway Bridge

Estimator: GAS

Cost Code	Work Type	Unit	Quantity	Direct Labor Cost	Labor Unit Cost	Equipment Cost	Equipment Unit Cost	Material Cost
	General Requirements							
01500.00	Move In	ls	1	$4,325	$4,325	$2,934	$2,934	$0
01700.00	Clean Up	ls	1	$2,142	$2,142	$1,104	$1,104	$0
	Subtotals			$6,467	$6,467	$4,038	$4,038	$0
	Sitework							
02220.10	Excavation, unclassified	cy	1,667	$1,308.00	$0.78	$619.00	$0.37	$0.00
02222.10	Excavation, structural	cy	120	$1,520.00	$12.67	$390.00	$3.25	$0.00
02226.10	Backfill, compacted	cu	340	$1,632.00	$4.80	$469.00	$1.38	$0.00
02350.00	Piledriving, rig mobilization & demobilization	ls	job	$3,506.00	$3,506.00	$5,508.00	$5,508.00	$200.00
02361.10	Piling, steel, driving	lf	2,240	$4,345.00	$1.94	$5,676.00	$2.53	$30,490.00
	Subtotals			$12,311.00	$3,526.19	$12,662.00	$5,515.53	$30,690.00
	Concrete							
03150.10	Footing forms, fabricate	sf	360	$666.00	$1.85	$0.00	$0.00	$193.00
03150.20	Abutment forms, prefabricate	sf	1,810	$1,991.00	$1.10	$0.00	$0.00	$1,645.00
03157.10	Footing forms, place	sf	720	$242.20	$0.34	$0.00	$0.00	$130.00
03159.10	Footing forms, strip	sf	720	$103.80	$0.14	$0.00	$0.00	$0.00
03157.20	Abutment forms, place	sf	3,620	$5,853.40	$1.62	$1,814.40	$0.50	$362.00
03159.20	Abutment forms, strip	sf	3,620	$2,508.60	$0.69	$777.60	$0.21	$0.00
03157.30	Deck forms, place	sf	1,800	$2,156.00	$1.20	$151.20	$0.08	$1,314.00
03159.30	Deck forms, strip	sf	1,800	$924.00	$0.51	$64.80	$0.04	$0.00
03251.10	Concrete, deck, saw joints	lf	60	$105.00	$1.75	$87.00	$1.45	$0.00
03311.10	Concrete, footings, place	cy	120	$611.00	$5.09	$580.00	$4.83	$6,237.00
03311.20	Concrete, abutments, place	cy	280	$4,654.00	$16.62	$3,430.00	$12.25	$14,553.00
03311.30	Concrete, deck, place & screed	sy	200	$820.00	$4.10	$696.00	$3.48	$2,911.00
03345.30	Concrete, deck, finish	sf	1,800	$1,638.00	$0.91	$44.00	$0.02	$342.00
03346.20	Concrete, abutments, rub	sf	1,960	$1,693.00	$0.86	$0.00	$0.00	$118.00
03370.20	Concrete, abutments, curing	sf	3,820	$110.00	$0.03	$0.00	$0.00	$69.00
03370.30	Concrete, deck, curing	sf	1,800	$54.00	$0.03	$0.00	$0.00	$36.00
	Subtotals			$24,130.00	$36.85	$7,645.00	$22.87	$27,910.00
	Metals							
05120.00	Steel, structural, place	lb	65,500	$1,758.00	$0.03	$1,728.00	$0.03	$27,052.00
05520.00	Guardrail	lf	120	$862.00	$7.18	$432.00	$3.60	$3,848.00
05812.00	Bearing plates	lb	3,200	$784.00	$0.25	$432.00	$0.14	$2,140.00
	Subtotals			$3,404.00	$7.46	$2,592.00	$3.76	$33,040.00

Figure 3.10 Highway bridge, project budget.

4

PROJECT
PLANNING

4.1 THE CPM PROCEDURE

Construction time control is a difficult, time-consuming, and arduous management function. Project managers work within an extremely complex and shifting time frame, and they need a management tool that will enable them to manipulate large numbers of job activities and complicated sequential relationships in a simple and understandable fashion. The Critical Path Method (CPM) is just such an expedient and constitutes the basis for the ensuing treatment of project time control. This method applies equally well to all construction work, large and small, intricate and straightforward.

The management techniques of CPM are based on a graphical project model called a network. This network presents in diagrammatic form those job activities that must be carried out and their mutual time dependencies. A diagram of this type is a simple and effective medium of communicating complex job interdependencies. It serves as a basis for the calculation of work schedules and provides a mechanism for controlling project time as the work progresses.

CPM is a three-phase procedure consisting of planning, scheduling, and time monitoring. Planning construction operations involves the determination of what must be done, how it is to be performed, and the sequential order in which it will be carried out. Scheduling determines calendar dates for the start and completion of project components. Time monitoring is the process of comparing actual job progress with the programmed schedule. Planning is the subject of this chapter. Scheduling and time monitoring are discussed in Chapters 5 and 9, respectively.

4.2 THE PLANNING PHASE

Planning is the devising of a workable scheme of operations that accomplishes an established objective when put into action. The most time-consuming and difficult aspect of the job-management system, planning, is also the most important. It requires an intimate knowledge of construction methods combined with the ability to visualize discrete work elements and to establish their mutual interdependencies. If planning were to be the only job analysis made, the time would be well spent. CPM planning involves a depth and thoroughness of study that gives the construction team an invaluable understanding and appreciation of job requirements.

Construction planning, as well as scheduling, must be done by people who are experienced in, and thoroughly familiar with, the type of field work involved. It is especially important that those who will be expected to implement the plan in the field have an opportunity to participate in its development. The project network and the management data obtained from it will be realistic and useful only if the job plan is produced and updated by those who understand the job to be done, the ways in which it can be accomplished, and the job-site conditions.

To construct the job network, input information from many sources must be sought. Guidance from key personnel involved with the project, such as estimators, project manager, site superintendent, and field engineer, can be obtained from a planning meeting, or perhaps a series of such get-togethers. The network serves as a medium whereby the job plan can be reviewed, criticized, modified, and improved. As problems arise, consultations with individuals can clear up specific questions. The important point here is that there should be full group participation in the development of the network and collective views must be solicited.

Participation by key subcontractors is also vital to the development of a workable plan. Normally, the prime contractor sets the general timing reference for the overall project. Individual subcontractors then review the portions of the plan relevant to their work and help to develop additional details pertaining to their operations. An important side effect is that this procedure brings subcontractors and the prime contractor together to discuss the project. Problems are detected early and steps toward their solutions are started well in advance.

It must be recognized that the project plan represents the best thinking available at the time it is conceived and implemented. However, no such scheme is ever perfect, and the need for changes is inevitable as the work goes along. Insight and greater job knowledge are acquired as the work progresses. This increased cognizance necessarily results in corrections, refinements, and improvements to the operational plan. The project program must be viewed as a dynamic device that is continuously modified to reflect the progressively more precise thinking of the field management team.

Construction planning may be said to consist of three steps: (1) determination of the job steps or "activities" that must be performed to construct the project, (2) ascertainment of the sequential relationships among these activities, and (3) the graphic presentation of this planning information in the form of a network. However, these three actions usually proceed more or less simultaneously with one another rather than as discrete and successive steps.

4.3 JOB ACTIVITIES

The segments into which a project is subdivided for planning purposes are called activities. An activity is a single work step that has a recognizable beginning and end and requires time for its accomplishment. The extent to which a project is subdivided depends on a number of practical considerations, but the following are suggested as guidelines for use when activities are being identified.

1. By area of responsibility. Work items done by the general contractor and each of its subcontractors should be separated.
2. By category of work as distinguished by craft or crew requirements.
3. By category of work as distinguished by equipment requirements.
4. By category of work as distinguished by materials such as concrete, timber, or steel.
5. By distinct structural elements such as footings, walls, beams, columns, or slabs.
6. By location on the project when different times or different crews will be involved.
7. With regard to owner's breakdown of the work for bidding or payment purposes.
8. With regard to the contractor's breakdown for estimating and cost-accounting purposes.

The activities used may represent relatively large segments of a project or may be limited to small steps. For example, a reinforced concrete wall may be only a single activity, or it may be broken down into erect outside forms, tie reinforcing steel, erect inside forms and bulkheads, pour concrete, strip forms, and cure. Trial and error, together with experience, are the best guides regarding the level of detail needed. What is suitable for one project may not be appropriate for another. Too little detail will limit planning and control effectiveness. Too much will inundate the project manager with voluminous data that tend to obscure the significant factors and needlessly increase the cost of the management system.

Not only is network detail a function of the individual project; it is also highly variable depending on who will be using the information. For example, the installation of a containment vessel on a nuclear power plant might be shown as a single activity by the project general contractor. However, the subcontractor responsible for moving and installing this enormously heavy vessel may require a complete planning network to accomplish the task. Planning detail also varies with the level of project management involved. This matter is further discussed in Sections 4.15 and 4.16.

4.4 JOB LOGIC

Job logic refers to the determined order in which the activities are to be accomplished in the field. The start of some activities obviously depends on the completion of others. A concrete wall cannot be poured until the forms are up and the reinforcing steel has been tied. Yet, many activities are independent of one another and can proceed concurrently. Much of job logic follows from well-established work sequences that are usual and standard in the trade. Nevertheless, for a project of any consequence, there is always more than one general approach and no unique order of procedure exists. It is the planner's responsibility to winnow the workable choices and select the most suitable alternatives. At times, this may require extensive studies, including the preparation of a separate network for each different approach to the work. Herein lies one of the major strengths of the CPM planning method. It is a versatile and powerful planning tool that is of great value in the time evaluation of alternative construction procedures.

Showing the job activities and their order of sequence (logic) in pictorial form produces the project network, this network being a graphical display of the proposed job plan. In general, job logic is not developed to any extent ahead of the network. Rather, job logic evolves in natural fashion as the job is discussed and the network diagram progresses. There are occasions, however, where it is useful to express limited areas of job logic in list form. In such instances, a listing that shows, for each activity, those succeeding activities that can start immediately after and only after that activity is finished is a sufficient and simple method of recording job logic.

4.5   RESTRAINTS

To be realistic, a job plan must reflect the practical restraints or limitations of one sort or another that apply to most job activities. Such restraints stem from a number of practical considerations. The restrictions of job logic itself might be called physical restraints. Placing forms and reinforcing steel might be thought of as restraints to pouring concrete. Such normal restrictions arise from necessary order in which construction operations are physically accomplished and are simply a part of job logic previously discussed. There are other kinds of restraints, however, that need to be recognized.

Reinforcing steel, for example, cannot be tied and placed until the steel is available on the site. Steel availability, however, depends on approval of shop drawings, steel fabrication, and its delivery to the job site. Consequently, the start of an activity involving the placing of reinforcing steel is restrained by the necessary preliminary actions of shop-drawing approval, steel fabrication, and delivery. Practical limitations of this sort on the start of some job activities are called "resource restraints," or more particularly in this case, "material restraints." Another common example of a job constraint is an "equipment restraint" where a given job activity cannot start until a certain piece of construction equipment becomes available. Other instances of job restraints are availability of special labor skills or crews, delivery of owner-provided materials, receipt of completed project design documents, and turnover of owner-provided sites or facilities. At times there may be "safety restraints," especially on the sequencing of structural operations on multistory buildings. The recognition and consideration of job restraints is an important part of job planning. Failure to consider such restraints can be disastrous to an otherwise adequate job plan.

Some restraints are shown as time-consuming activities. For example, the preparation of shop drawings and the fabrication and delivery of job materials are material restraints that require time to accomplish and are depicted as activities on project networks. Restraints are also shown in the form of dependencies between activities. If the same crane is required by two activities, the equipment restraint is imposed by having the start of one activity depend on the finish of the other.

4.6 USE OF PROJECT OUTLINE

The initial stages of planning are often accomplished by making a project outline. This outline can be made manually or with the use of an outliner in a computer word processing program. Regardless of how large or complex a project may be, it can be readily broken down into its major components. These principal job elements form the first level of the outline. In the case of the highway bridge, the major components are:

Procurement
Field mobilization & site work
Pile foundations
Concrete abutments
Deck
Finishing operations

In turn, each of these major project segments can be broken down into its subcomponents to form a second outline level. For example:

Concrete abutments
    Abutment #1
    Abutment #2

Similarly, each of these can be further sub-divided.

Abutment #1
    Forms and rebar abutment #1
    Pour abutment #1
    Strip & cure abutment #1
    Backfill abutment #1
    Rub concrete abutment #1

A complete outline is shown in Appendix B. This outline is an example of a work breakdown structure and has a number of other applications in addition to the planning function. In Chapter 9 the outline will be used as the basis of a hierarchy of reports and bar charts and in Chapter 10 it will be used for cost reporting.

This outline planning process can be performed by the project management team. Brainstorming works very well in this process, and the preparation of the outline can move along rapidly. As can be seen from the example just cited, the outline develops quite naturally from basic project components through a number of levels to activity-size segments of the project.

When the outline is reasonably complete, the arrangement of the activities and their logical dependencies constitute the next step. Here again, the project management team can determine the necessary order of activity accomplishment. It is during this type of planning meeting that the project management team begins to visualize the project as a whole and the manner in which the individual pieces fit together. One team member may suggest that one of the abutments be backfilled before the deck girders are delivered so that the crane can unload the girders from the truck and put them directly into place. Another member might propose that both abutments be backfilled before the deck concreting operations start so that the concrete trucks can reach both ends of the bridge deck. This would enable the concrete to be placed directly from the trucks. The person drawing the network, probably with the aid of computer graphics and a computer projector, will record each of the suggestions made. In such a process, the planning team can continually review the information and make informed decisions regarding the project construction plan. In addition to the generation of the planning information, the project management group begins to think and act as a team.

4.7 PRECEDENCE NOTATION

There are two symbolic conventions currently used to draw construction networks. One depicts each activity as a rectangular box, this being called "precedence" or "activity-on-node" notation. The other shows each activity as an arrow and is called "arrow" or "activity-on-arrow" notation. Precedence diagrams have several important advantages over arrow diagrams, a topic that is discussed in Section 6.19. Because of these advantages, precedence notation will be emphasized in this text. However, because the usage of arrow diagrams is not uncommon, the basics of planning and scheduling using arrow notation are also discussed.

In drawing a precedence network, each time-consuming activity is portrayed by a rectangular figure. The dependencies between activities are indicated by dependency or sequence lines going from one activity to another. The identity of the activity and a considerable amount of other information pertaining to it are entered into its rectangular box. This matter will be further developed as the discussion progresses.

4.8 THE PRECEDENCE DIAGRAM

The preparation of a realistic precedence diagram requires time, effort, and experience with the type of construction involved. The management data extracted from the network can be no better than the diagram itself. The diagram is the key to the entire time-control process. When the network is first being developed, the planner must concentrate on job logic. The only consideration, at this stage, is to establish a complete and accurate picture of activity dependencies and inter-relationships. Restraints that can be recognized at this point should be included. The time durations of the individual activities are not of concern during the planning stage. Additionally, it is assumed that the labor and equipment needs of the activities can be met as they arise, other than where advance recognition of restraints or resource conflicts is possible. Matters of activity times and resource conflicts will be considered in detail later during project scheduling.

To illustrate the mechanics of network diagramming, consider a simple project such as the construction of a pile-supported concrete footing. The sequence of operations will be the following. Excavation, building of footing forms, and the procurement of piles and reinforcing steel are all opening activities that can begin immediately and can proceed independently of one another. After the excavation has been completed and the piles are delivered, pile driving can start. Fine grading will follow the driving of piles, but the forms cannot be set until the piles are driven and the forms have been built. The reinforcing steel cannot be placed until fine grading, form setting, and rebar procurement have all been carried to completion. Concrete will be poured after the rebar has been placed, and stripping of the forms will terminate this sequence. In elementary form, this is the kind of information generated as a project is analyzed by the planning group. The precedence diagram describing the prescribed sequence of activities is shown in Figure 4.1.

Each activity in the network must be preceded either by the start of the project or by the completion of a previous activity. Each path through the network must be continuous with no gaps, discontinuities, or dangling activities. Consequently, all activities must have at least one activity following, except the activity that terminates the project. It is standard practice, as well as being a requirement of some computer programs, that precedence diagrams start with a single opening activity and conclude with a single closing activity. In this regard, "Start" and "Finish" activities appear in precedence diagrams throughout this text. These two zero-time figures are shown herein with rounded corners to distinguish them from the usual time consuming activities.

The numbering of network activities, as shown in Figure 4.1, is not standard practice although it is used in this text for the purpose of easy and convenient activity identification. When network activities are numbered, each activity should have a unique numerical designation with the numbering proceeding generally from project start to project finish. Usual practice is that numbering is not done until after the network has been completed. Leaving gaps in the activity numbers is desirable so that spare numbers are available for subsequent refinements and revisions. In this text, the network activities are given identifying names and are numbered by multiples of 5 or 10.

4.9 NETWORK FORMAT

A horizontal diagram format has become standard in the construction industry. The general synthesis of a network is from start to finish, from project beginning on the left to project completion on the right. The sequential relationship of one activity to

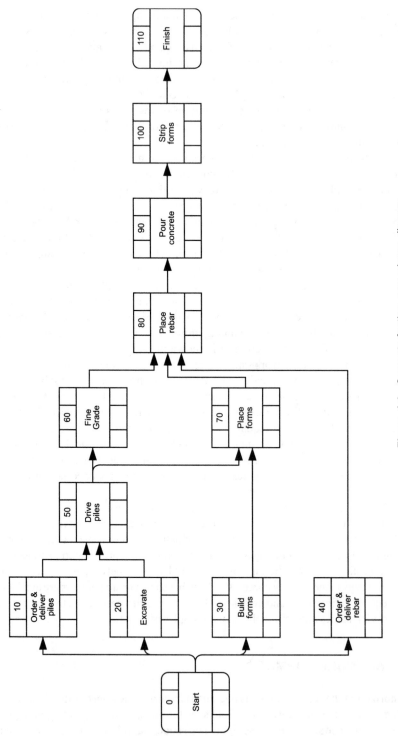

Figure 4.1 Concrete footing, precedence diagram.

another is shown by the dependency lines between them and the essence of the network is the manner in which the constituent activities are joined together into a total operational pattern. In the usual precedence diagram, the length of the lines between activities has no significance because they indicate only the dependency of one activity on another. Arrowheads are not always shown on the dependency lines because of the obvious left to right flow of time. However, arrowheads are shown herein on precedence diagrams for additional clarity.

During initial diagram development, the network is sketched emphasizing activity relationships rather than the appearance or style of the diagram. Corrections and revisions are plentiful. The rough and ready appearance of the first version of the diagram is of no concern; its completeness and accuracy are. The finished diagram can always be put into a more tidy form at a later date.

The use of symbols to identify activities rather than writing out the description of each activity is not recommended. For instance, "PR" might be used for "Place rebar" or "PC" for "Pour concrete." Even mnemonic codes such as this make a diagram difficult to read, however, and the user must spend a considerable amount of time consulting the symbol listing to check the identity of activities. Networks are much more intelligible and useful when each activity is clearly identified on the diagram.

When a working precedence diagram is being prepared, the scale and spacing of the activities deserve attention. If the scale is too big and activities are widely spaced, the resulting network is likely to become so large that it is unmanageable. On the other hand, a small scale and overly compact makeup render the diagram difficult to read and inhibit corrections and modifications. With experience and observation of the work of others, practitioners will soon learn to adjust the scale and structure of their diagrams to the scope and complexity of the project involved. It must not be forgotten that the network is intended to be an everyday tool, used and consulted by a variety of people. Emphasis here is not on drafting elegance, but on contriving the most realistic and intelligible form of network possible.

Dependency lines that go backward from one activity to another should not be used and are not possible with most computer programs. Backward, in this instance, means going from right to left on the diagram which is against the established direction of time flow. Backward sequence lines are confusing and increase the chances of unintentional logical loops being included in the network. A logical loop involves the impossible requirement that an activity be followed by an activity that has already been accomplished. Logical loops are, of course, completely illogical, but they can be inadvertently included in large and complex networks if backwardly directed sequence lines are permitted. Crossovers occur when one dependency line must cross over another to satisfy job logic. Careful layout will minimize the number of crossovers, but there are usually some that cannot be avoided.

Color coding the diagram can be very useful at times. Various colors can be used to indicate different trades, work classifications, major job segments, and work that is subcontracted.

4.10   LAG RELATIONSHIPS

Figure 4.1 has been drawn on the customary basis that a given activity cannot start until <u>all</u> of those activities immediately preceding it have been completed. Also inherent in the notation used in Figure 4.1 is that an activity can start once all of its immediately preceding activities have been finished. In the figure, activity 60, "Fine grade," cannot start until activity 50, "Drive piles," has been finished, and the start of activity 90, "Pour concrete," must await completion of activity 80, "Place rebar." In addition, by way of example, activity 80, "Place rebar," can start immediately once activity 60, "Fine grade," activity 70, "Place forms," and activity 40, "Order and deliver rebar," have all been finished. Under these conditions, it is not possible to have the finish of one activity overlap the start of a following activity. Where such a condition potentially exists, the work must be further subdivided.

There are cases, however, where there may be a delay between the completion of one activity and the start of a following activity, or there is a need to show that one activity will overlap another in some fashion. Precedence diagrams can be made to show a variety of such conditions using lag relationships. This concept of lags is developed more completely in Section 5.20.

4.11   PRECEDENCE DIAGRAM FOR HIGHWAY BRIDGE

When first beginning the development of a job network, it is wise to minimize detail, especially if the project is substantial in extent. Getting started on a large project can be overwhelming and a general job plan of limited size can be useful as a means of getting started. Once a general job plan has been put together, it is an excellent framework on which to amplify the network to the level of detail desired. Some practitioners favor starting job planning by compiling a relatively short list of major project operations arranged in chronological order. On the highway bridge, previously described in Section 3.4, this initial list could be the following:

Excavation
Piles, abutment #1
Piles, abutment #2
Abutment #1
Abutment #2
Steel girders
Deck slab
Final operations

This list can be used to prepare a preliminary job plan such as the one shown in Figure 4.2. The major operations in the preceding list are much the same as, and could be identical with those previously identified by the bar chart in Figure 3.5 when a preliminary time analysis of the project was made during the estimating period. A general job plan like that shown in Figure 4.2 is useful in the sense that it places the entire project in perspective. It is profitable for the planner to establish a general frame of reference before he begins to struggle with the intricacies of detailed job planning.

All of the steps to job planning have now been reviewed and, providing that the simple rules of precedence diagramming are followed, it should be possible to develop a working diagram for the highway bridge. A company prebid conference has established the following general ground rules of procedure. One abutment at a time will be constructed for reasons of equipment limitations and form-material economy. Abutment #1 will be constructed first with its footing being poured initially, followed by the breast and wing walls. As excavation, pile driving, forming and pouring are completed on abutment #1, these operations move over to abutment #2. Only one set of footing forms and abutment wall forms will be made, these being used first on abutment #1 and then moved over to abutment #2.

As soon as the steel girders have been delivered, the concrete abutments stripped, and abutment #1 backfilled, the steel girders will be placed. Forms and reinforcing steel for the deck slab can follow. Abutment #2 must be backfilled before the deck can be poured. Concrete rubbing can be started as the abutments are stripped and must be finished before painting can start. Job cleanup and final inspection complete the project.

Figure 4.3 is the precedence diagram that results from the highway bridge logic just established. Most of the dependencies on the diagram are normal dependencies in that they are the natural result of the physical nature of the activities themselves. However, Figure 4.3 also includes some resource restraints. When such restraints have been firmly established, it is advisable to include them in the first draft of the network because they can be significant planning factors that have a major effect upon the project plan and schedule. By way of explanation, refer to activities 10, 20, and 50 in Figure 4.3, using activity 20 as a typical example. Activity 20, "Prepare & approve S/D, footing rebar," is a consequence of the usual construction contract requirement that the prime contractor and its subcontractors must submit shop drawings (S/D) and other descriptive information concerning project materials and machinery to the owner or project design professional for their approval. This must be done before such job requirements can be provided to the job site and used thereon. Shop drawing approval and material fabrication and delivery are shown separately on Figure 4.3 because the delivery of reinforcing steel and other materials to the project is usually quoted by the vendor as requiring a stipulated period of time after return to the vendor by the general contractor of approved shop drawings.

With regard to material restraints such as activity 70, an additional matter may have to be considered. For example, if the footing rebar is sold to the contractor "FOB trucks, job site," and this is very often the case, the contractor is responsible for unloading the vendor's trucks when they arrive on the job site. In such an instance, the planner may wish to advertise this fact by including another activity, "Unload foot. rebar," at the end of activity 70. This simply serves as a reminder to all concerned that the contractor must make advance arrangements to have suitable unloading equipment and men available when delivery is made. Unloading activities have not been included in Figure 4.3, however, so that the discussion can concentrate on basics.

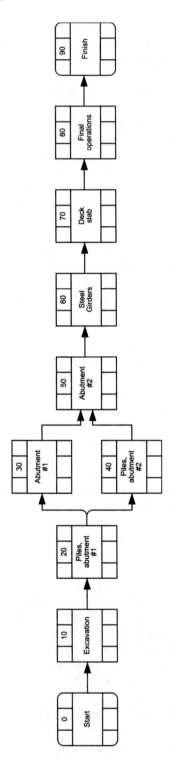

Figure 4.2 Highway bridge, general job plan.

An equipment restraint has also been included in Figure 4.3. The driving of steel pilings for the two abutments requires an assembly of equipment units, including a large crane, that must be mobilized before pile driving can start. This is illustrated by activity 100, "Mobilize pile driving rig," that precedes activity 110, "Drive piles abutment #1." After the piles are driven for both abutments, the pile driving rig is demobilized. As shown by Figure 4.3, this disassembly process must occur before activity 180, "Forms & rebar abutment #1," can start because the crane used for pile driving is required to handle the forms and rebar for this abutment. Abutment #2 follows at a later date. Another resource constraint in Figure 4.3 arises because the same concrete forms are going to be used for both abutments. As a result of this decision, activity 240, "Forms and rebar, abutment #2," cannot start until activity 220, "Strip and cure abutment #1," has been completed. In a similar manner, the same forms will be used for both footings.

4.12 VALUE OF PRECEDENCE NETWORK

Figure 4.3 is not only a lucid job model, it is also an effective tool useful for day-to-day direction and control of the work. Figure 4.3 *is* the job plan for the highway bridge project. Preparing the network has forced the job planners to think the job through completely from start to finish. Decisions have been made about equipment, construction methods, and sequence of operations. The field supervisory team now possesses a depth of knowledge about the project that can be obtained only through such a disciplined process of detailed job analysis.

The job plan, in the form of a precedence diagram, is comprehensive, detailed, and in a form that is easy to impart to others. The network diagram is an expedient medium for communication between field and office forces. If, for some reason, the project manager or field superintendent must be changed during construction, the diagram can assist appreciably in effecting a smooth transition. The diagram makes job coordination with material dealers, subcontractors, owners, and architect-engineers a much easier matter. The orderly approach and analytical thinking that have gone into the network diagram have produced a job plan far superior to any form of bar chart or narrative analysis. Invariably, synthesis of the network results in improvements to original ideas and a sharpening of the entire approach to the project.

4.13 REPETITIVE OPERATIONS

Some kinds of construction projects involve long series of repetitive operations. Pole lines, highways, pipelines, multistory buildings, and tract housing are familiar examples of construction jobs that entail several parallel strings of continuing operations. One segment of the Example Project described in Section 2.17 is the relocation of 5 miles of natural gas pipeline. This is a good example of a construction operation that will involve repetitive operations that proceed simultaneously with one another.

For purposes of discussing the planning of such a project, the major repetitive segments of the pipeline relocation have been identified as locate and clear, excavate, string pipe, lay pipe, test, and backfill. The basic plan for this job is shown in Figure 4.4 with all the operations being sequential except for excavate and string pipe, which are done concurrently. However, no pipeline contractor would proceed in such a single-file manner unless the pipeline was very short in length or some special circumstance applied. Rather, location and clearing work would get well underway. Excavation, together with pipe stringing, would then start and pipe laying would proceed fairly closely behind. Pressure testing of the pipe and backfilling complete the sequence. After the project gets "strung out" along the right-of-way, all these operations will be moving ahead simultaneously, one stage following the next.

All this makes it clear that more detail than is included in Figure 4.4 will be necessary if the job plan is to be useful. The way to accomplish this is to divide the pipeline into arbitrary but typical repeating sections. The length of repeating section chosen can be quite variable, depending on the length of the pipeline, terrain, contract provisions, and other job conditions. For discussion purposes, suppose that the contractor decides that a mile-long section of pipeline represents a fairly typical unit of work. This would be exclusive of river, railroad, and highway crossings which are frequently done in advance of the main pipeline and which may require their own planning and scheduling studies as separate operations. Figure 4.5 indicates how the basic plan can be broken down into mile-long units.

The notation used in Figure 4.5 is one where the individual activity box only indicates the section of the pipeline involved. The work categories listed at the left margin of the figure apply to the horizontal string of activities at each successive level. The logic of Figure 4.5 shows that after a mile of the pipeline right-of-way has been located and cleared, both excavate and pipe stringing start. These latter two activities proceed simultaneously, but one does not depend on the progress of the other. After excavation and string pipe have proceeded 1 mile, pipe laying starts. Testing and backfill begin in the order shown. Once a work phase is started, it will proceed more or less continuously until its completion.

4.14 NETWORK INTERFACES

Often different portions of the same project are planned separately from one another. However, as frequently happens, the individual construction plans are not truly independent of one another and the two networks are actually related in some way. Thus the two networks must "interface" with each other so that mutual dependencies are properly transmitted from one plan to the other. An interface refers to the dependency between activities of two different networks. This dependency can be indicated on the networks by interrupted sequence lines between the affected activities.

The pipeline relocation will be used to illustrate the workings of an interface. Assume that this job includes a stream crossing at the end of the third mile of pipeline. The construction of the crossing structure will be separate from that of the pipeline, but the two must be correlated so the crossing structure is ready for pipe by the time the pipeline construction has progressed to the crossing location. In this instance, planning of the pipeline itself and of the crossing structure proceed independently of one another. Figure 4.5 is the planning network for the pipeline work. The crossing is to be effected by a suspension-type structure, and Figure 4.6 is its construction plan.

The required job logic of having the crossing structure ready to accept pipe by the time the pipeline has reached the crossing site can be imposed by having activity 110 of Figure 4.6 immediately follow activity 170 of Figure 4.5. This is shown in both figures by the dashed dependency line. Ensuring that the crossing structure is started soon enough for the desired meshing of the two networks is now a matter of scheduling and will be treated in Chapter 5.

4.15   THE MASTER NETWORK

As illustrated by Figure 2.1, the total Example Project consists of several major subprojects, each being relatively self-contained and independent of the others. The planning networks of two of these segments of the Example Project, the highway bridge and the pipeline relocation, have been developed. It is now easy to visualize that similar planning diagrams must be developed for the river diversion, haul roads, borrow development, earth dam, and other major project elements. These separate networks would be used for the detailed time scheduling and daily field management of the several components of the Example Project. In a general sense, each of the major project segments is constructed and managed separately.

The planning for each subproject is done in considerable detail. This fact is obvious from the nature of the networks developed for the highway bridge and the pipeline relocation. Although this level of detail is necessary for the day-to-day field control of construction operations, it is overwhelming for an owner or project manager who wants to keep abreast of overall site operations in a more general way. The amount of detail required by managers decreases with the span of their authority. A field manager requires information and data that are tailored to the level of his responsibilities.

On the Example Project, a master planning network would be prepared that encompasses and includes every subproject. The level of detail of this diagram would be relatively gross, with each activity representing substantial segments of the field work. The highway bridge and the pipeline relocation would each appear as a small cluster of associated activities. It is likely that the highway bridge would appear in the master diagram as the activities shown in Figure 4.2, and the pipeline relocation would be as presented in Figure 4.4. The master job plan, of necessity, concerns itself essentially with the big picture, the broad aspects of the major job segments and how they relate to each other. Keeping the master network free of excessive detail is necessary to the production of an overall plan that can be comprehended and implemented by those who must apply it.

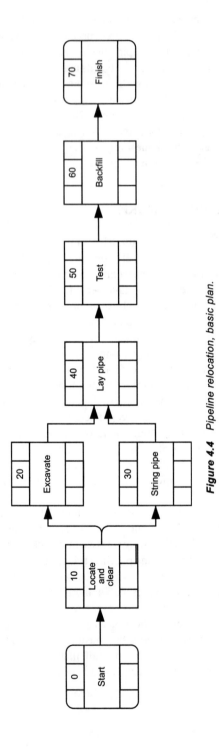

Figure 4.4   *Pipeline relocation, basic plan.*

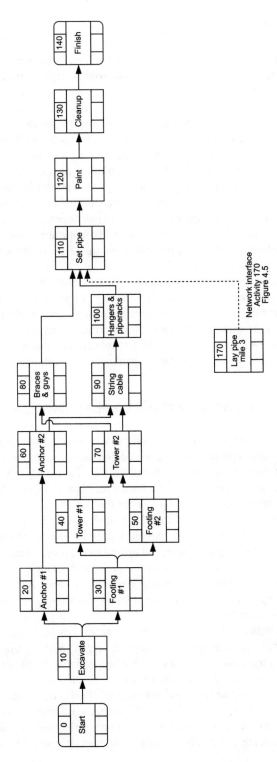

Figure 4.6   *Pipeline crossing structure, precedence diagram planning.*

4.16 SUBNETWORKS

The detailed planning diagrams of the highway bridge, Figure 4.3, and of the pipeline relocation, Figure 4.5, are spoken of as "subnetworks" of the master network. As has been discussed, they are used by lower-level field management for everyday direction and control of the work. It would be possible, of course, to draw an overall project network that would combine the detailed planning networks of all the subprojects into one enormous diagram. However, such an all-inclusive network would be so large and so cumbersome as to be virtually useless to anybody concerned with the Example Project. The detailed planning of an extensive construction contract such as the Example Project is accomplished primarily through the medium of many subnetworks, with their interdependencies being indicated by appropriate interfaces. In this way, management personnel can concentrate on their own local-ized operations. Each manager can monitor his work according to his own plan without the distraction of having to wade through masses of information irrelevant to him and his responsibilities.

It is also entirely possible that certain activities of the highway bridge network (Figure 4.3) or of the pipeline relocation network (Figure 4.5) might require further expansion into more detailed subnetworks. To ensure their timely accomplishment, it is sometimes desirable to subject certain critical activities to further detailed planning study. To illustrate, activity 100, "Mobilize pile driving rig," of Figure 4.3 might be expanded into its own planning subnetwork as shown by Figure 4.7.

4.17 COMPUTER APPLICATIONS TO PLANNING

Computers are having a major impact on the way construction professionals plan, schedule, and control projects. In the succeeding chapters, computer applications to each phase of construction management will be discussed. Because of the rapid changes in computer technology and software sophistication, only the fundamentals of computer applications are appropriate for discussion.

Earlier in this chapter it was stated that planning is the most important phase of project management and it is also the most difficult and time consuming. This is because the knowledge, experience, and insight of the project team must be brought together to identify a plan which is, at the same time, complex and uncertain. For years, the process was one of planning meetings, drawing the network on paper, another planning meeting, revising the network drawing again and again until the team was satisfied or ran out of time.

Computers with graphics capabilities have made a major impact on the planning process. Networks are drawn on the computer screen. With a projection device, the network is displayed on a screen as it is being developed by the project team. One person at the computer records each idea as the plan is being developed. Each addition to the plan can be seen, understood, and, if necessary, criticized by the other team members. In this way the skills of the team are combined and a dynamic plan is developed in much less time than previously possible. Networks developed on the computer screen are extremely flexible. Activities or groups of activities can be

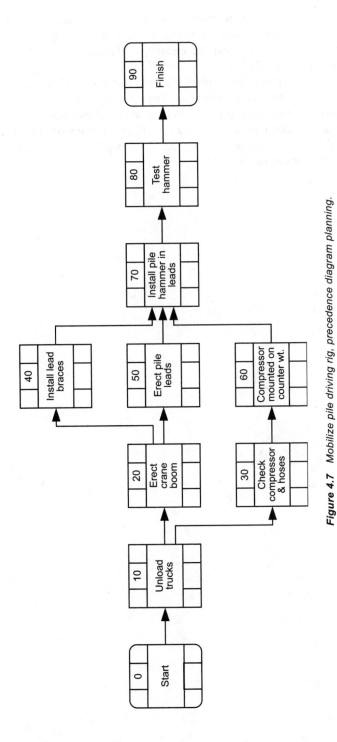

Figure 4.7 Mobilize pile driving rig, precedence diagram planning.

moved around on the screen. Individual parts of the network can be created independently and then combined with the main network. Multiple approaches to a particular planning problem can be created, evaluated, compared and the best solution used in the final plan.

Perhaps the best part of this type of planning is that it involves each member of the project team in the planning process. The resulting plan is perceived to be the team's plan rather than a plan provided by the planning department. Everyone on the team now has a stake in making this plan work. This perception alone makes each phase of the project management system described in the succeeding chapters contribute to a successful project.

5

PROJECT
SCHEDULING

5.1 SCHEDULING PROCEDURE

Chapter 4 dealt with the procedures followed in developing the planning diagram for a construction project. Once the network diagram has been developed, the time-management system enters into a new phase, that of work scheduling. At this stage a new element, time, is introduced into the planning diagram. On no occasion during the development of the job plan was time taken into account, not with regard to the overall construction period required nor the times necessary for completion of the individual activities. This chapter is concerned with the time scheduling of construction projects.

As has been stated previously, the Critical Path Method (CPM) was developed especially for the planning and scheduling of construction operations and is the procedure used throughout this text. However, it is of interest to note that a somewhat different procedure, Program Evaluation and Review Technique (PERT), has been devised for application to the scheduling of research and development projects. In such cases, there is little or no background of past experience with similar undertakings as there is with construction and time estimates of reasonable accuracy are very difficult to make. Although basically similar, CPM and PERT differ in several important respects with regard to the making of activity time estimates. Some knowledge of the PERT procedure can broaden the construction manager's perspective and is discussed in Appendix C. In this chapter, the CPM scheduling method will now be developed and applied.

A project schedule is a projected timetable of construction operations that will serve as the principal guideline for project execution. There are several steps involved in the devising of an efficient and workable job schedule. The following list is offered as a procedural guide in this regard.

1. Estimate the time required to carry out each network activity.
2. Using these time estimates, compute the time period required for overall project completion.
3. Establish time intervals within which each activity must start and finish to satisfy the completion date requirement.
4. Identify those activities whose expedient execution is crucial to timely project completion.
5. If the project completion date is not consonant with contract or other requirements, shorten the project duration at least possible cost.
6. Using surplus or float times that most activities possess, adjust the start and finish times of selected activities to minimize resource conflicts and smooth out demands for manpower and equipment.
7. Make up a working project schedule that shows anticipated calendar dates for the start and finish of each network activity.

This chapter discusses the first four steps just described. The remainder are presented in subsequent chapters. The highway bridge and pipeline relocation will be used in conjunction with the presentation of the several facets of construction scheduling.

5.2 ACTIVITY TIMES

CPM customarily uses a single time estimate for each network activity. This is because, in construction, each activity is "deterministic" in the sense that similar or identical work has been performed many times before. Such prior experience enables the contractor to estimate with reasonable accuracy the time required to carry out each job operation. Single time estimates for network activities will be used exclusively in the main body of this book.

In the construction industry, activity durations are customarily expressed in terms of working days, although other time units such as hours, shifts, or calendar weeks are sometimes used. The main criterion is that the unit chosen be harmonious with the project and the management procedures being used. Once a time unit is selected, it must be used consistently throughout the network. More will be said about this later in the chapter. Working days will be used herein as the standard time unit for activity duration estimates and scheduling computations.

5.3 RULES FOR ESTIMATING ACTIVITY DURATIONS

The true worth of a project work schedule and the confidence with which it can be applied depend almost entirely on the job logic as contained in the planning network and the accuracy with which the individual activity times can be estimated. As a general rule, time estimates can be made more reliably for activities of limited scope than for those of larger extent. In itself, this is not necessarily an argument for the use of great detail in job planning. Nevertheless, when considerable uncertainty surrounds the duration of a given activity, it can sometimes be helpful to subdivide that activity into smaller elements.

The following important rules apply to the estimation of activity durations:

1. Evaluate activities one at a time, independently of all others. For a given activity, assume that materials, labor, equipment, and other needs will be available when required. If there is reason to believe that this will not be true, then the use of a preceding restraint may be in order.

2. For each activity, assume a normal level of manpower and/or equipment. Exactly what "normal" is in this context is difficult to define. Most activities require only a single crew of workers or a standard spread of equipment. Based on experience, conventional crew sizes and equipment spreads have emerged as being efficient and economical. In short, a normal level is about optimum insofar as expedient completion and minimum costs are concerned. A normal level may be dictated by the availability of labor and equipment. If shortages are anticipated, this factor must be taken into account. However, conflicting demands among concurrent activities for workers or equipment are ignored for the time being. At this stage, such conflicts are only matters of conjecture, and they will be investigated in detail during a later stage of scheduling.

3. If time units of working days are being used, assume a normal workday. Do not consider overtime or multiple shifts unless this is a usual procedure or a part of the standard workday. Round-the-clock operations are normal in most tunnel work, for example, and overtime is extensively used on highway jobs during the summer months to beat the approaching cold weather. Some labor contracts guarantee overtime work as a part of the usual workday or workweek. In these cases, the extra hours are normal and should be considered.

4. Concentrate on estimating the duration of the individual activity and ignore all other time considerations. In particular, the completion date of the project must be put entirely out of mind. Otherwise, there is apt to be an effort made, consciously or unconsciously, to fit the activities within the total time available. This is one of the serious drawbacks of the bar chart as a planning and scheduling device. Most contractors will admit that the average bar chart is made up primarily by adjusting the individual work items to fit within an overall time requirement. The only consideration pertinent to estimating an activity duration is how much time is required to accomplish *that* activity, and that activity alone.

5. Use consistent time units throughout. It must be remembered that when using the working day as a time unit, weekends and holidays are not included. Certain job activities such as concrete curing or systems testing carry through nonworking days. To some extent, allowances can be made for this. If curing periods of seven days are involved, for example, this will involve only five working days. In most cases, however, such corrections cannot be exact and

must be based on the scheduler's best judgment. Also associated with the use of consistent time units is the matter of conversion of time periods from one base to another. For example, vendors invariably express delivery times in terms of calendar days. If the delivery of a pump is given as 30 days by the vendor, this will translate into approximately 21 working days. Most computer programs handle these problems with multiple calendars. Manual calculations require the scheduler to make these adjustments to the durations.

5.4 ESTIMATING ACTIVITY DURATIONS

It is important that someone experienced in and familiar with the type of work involved be consulted when the activity durations are being estimated. With respect to work done by subcontractors, it is good practice to solicit input from them concerning the times required by those activities that are their responsibility. The subcontractor is usually in the best position to render judgments concerning the times required for the accomplishment of its work.

One effective way of estimating an activity duration is to compute it by applying a crew or equipment production rate to the total number of units of work to be done. For illustrative purposes, the determination of time estimates for two of the activities on the highway bridge will be discussed. First, consider activity 110, "Drive piles, abutment #1," as it appears in Figure 4.3. The summary sheet for Bid Item 4 in Appendix A shows that a pile-driving production rate of 70 linear feet per hour was used when the cost of the highway bridge was estimated. Each abutment involves the driving of twenty-eight 40-foot long pilings. Dividing the total lineal footage of 1120 feet by 70 gives 16 hours or two working days. In addition to the actual pile driving, one day will be required to prepare the templates, pile head cutoff, and moving the equipment. Thus, the time estimate for this particular activity will be three days. In a similar manner for activity 200, "Pour abutment #1," Figure 3.6 shows that the concrete for this abutment will be poured at the rate of 10 cubic yards per hour. This production rate can be used to compute the time required to pour the abutment. Each abutment contains 140 cubic yards of concrete and dividing this by 10 cubic yards per hour gives 14 hours or about two working days.

Another approach in determining activity times is to assume a crew size and use the estimated labor unit cost rather than a production rate. To illustrate this procedure, Bid Item 3 in Appendix A shows the unit labor cost for compacted backfill to be $4.80 per cubic yard. Activities 280 and 310 each include 170 cubic yards of compacted backfill. Thus, each of these activities has a direct labor cost of 170($4.80) = $816. Assume a crew of three laborers with a daily labor cost of 8(3)($11.60) = $278.40. Dividing $816 by $278.40 gives 2.93 or three days. Hence, the estimated duration of activities 280 and 310 will each be three working days.

When estimating activity times, one special circumstance must be kept in mind. This is the case where the same work item is repeated several times during the construction period. For example, successive job activities may involve repetitions of essentially identical concrete forming. The time performance on such work will improve considerably during the first few cycles. This is a "learning curve" effect that causes the later items to be accomplished in less time than were the first ones. The basic proposition of the learning curve phenomenon is that skill and productivity in performing the same work improves with experience and practice and should be reflected in the network durations.

Time estimates of surprising accuracy can often be made informally. Experienced construction supervisors have an almost uncanny ability to give off-the-cuff time estimates that usually prove to be reasonably close. This may seem to be an almost casual approach to such an important matter, but experience shows that it has its place, especially in the checking of time values obtained by presumably more exact means. Input from field superintendents is valuable and desirable, but it would be a mistake to allow them to make all the duration estimates themselves in an informal fashion. If for no other reason, they are human and their time estimates are apt to be generous so that their chances of staying on schedule later on are commensurately improved.

Activity durations are customarily expressed in terms of full working days because, in most cases, to do otherwise is to assume a fictitious degree of accuracy. If an activity time is less than one working day, the activity concerned may be too small for practical job scheduling and control. Figure 5.1, the logic for which has been previously presented in Figure 4.3, shows the estimated durations for each activity of the highway bridge. Inspection of the sample activity shown in Figure 5.1 discloses that each activity duration in terms of working days is shown in the lower, central part of the activity box. Each activity is also given an identifying number, this being located in the upper, central part of the activity box. The other numerical values shown with the activities and the contingency activity at the right end of Figure 5.1 will be discussed subsequently.

5.5 TIME CONTINGENCY

When applying the CPM procedure to a construction project, it is assumed that the individual activity durations are deterministic in the sense that they can be estimated relatively accurately and that their actual durations will have only relatively minor variances from the estimated values. An estimated activity duration is the time required for its usual accomplishment and does not include any allowance for random or unusual happenings. Using such estimates of individual activity times, an estimated or likely completion time for the entire project can be computed. The actual project completion time will probably vary from this estimated or mean value. These possible values of actual completion time distribute themselves in a symmetrical pattern about the estimated value in a symmetrical pattern of possible under-run or over-run completion time durations, this pattern constituting a normal probability distribution. See Appendix C for more information regarding variations in completion dates.

Thus far, the assumption has been that the activity times account entirely for the project completion period. When the time estimate is made for an activity, it is based on the assumption that "normal" conditions will prevail during its accomplishment. Although normal conditions are difficult to define, it is a concise enough concept to recognize that there are many possibilities of "abnormal" occurrences that can substantially increase the construction time that will actually be required. Allowances for abnormal or random delays are accounted for in the form of a contingency allowance.

Different procedures are followed with respect to adding contingency allowances to project durations to account for the occasional losses of time that will probably occur during the construction process. A contingency allowance cannot usually be applied to individual activities to account for general project delays such as those caused by fires, accidents, equipment breakdowns, labor problems, late material deliveries, bad weather, damage to material shipments, unanticipated site difficulties, and the like. It is impossible, in almost all cases, to predict which activities may be affected and by how much. As a result, a general allowance for such time contingencies is normally added to the overall project duration.

5.6 PROJECT WEATHER DELAYS

Probably the most common example of project delay is that caused by inclement weather. It is important that the probable effect of adverse weather be reflected in the final project time schedule. The usual basis for making time estimates is on the assumption that construction operations will proceed on every working day. However, it is obvious that there can be a profound difference in the time required for excavation depending on whether the work is to be accomplished during dry or wet weather. In similar fashion, snow, cold temperatures, and high winds can substantially affect the times required to do certain types of construction work.

How allowances for time lost because of inclement weather are handled depends to a great extent on the type of work involved. When projects, such as most highway, heavy, and utility work, are affected by the weather, the entire project is normally shut down. As a result, it is usual that an estimated number of days lost because of weather be added to the overall project duration.

On buildings and other work that can be protected from the weather, allowances for time lost are commonly applied to those particular activities susceptible to weather delay. It is not often that a job of this type is completely shut down by bad weather after it is enclosed. Although some parts of the job may be at a standstill, others can still proceed. Consequently, a better overall job of scheduling will result if allowances for weather are ascribed to those groups of activities likely to be involved rather than adding a weather contingency to the entire project.

The approximate time schedule in Figure 3.5 that was devised to assist with estimating project cost can provide general information concerning the seasons of the year during which various portions of work will be done. When activity time durations are being estimated, a check can be made using this schedule to determine where weather delays are likely and what activities will be adversely affected. One method of taking weather delay into account is to add a time contingency to the normal durations of specific activities that are deemed to be weather sensitive. Probably a better procedure in this regard is to include, wherever needed, a weather contingency activity in the network to reflect the probable time loss associated with an activity or string of activities. A contingency activity in a network is labeled as such and is treated, for scheduling purposes, just like any regular activity. In the ensuing discussion of scheduling the highway bridge, no time contingencies will be added to individual activities. The only allowance for lost time will be that added to overall project duration in the form of a terminal contingency activity. This is discussed in Section 5.9.

5.7   NETWORK TIME COMPUTATIONS

The determination of a detailed activity time schedule is normally done by computer, these times being expressed in terms of calendar dates or expired working days. However, the optimum use of this information for project time control purposes requires that the user have a thorough understanding of the computations and the true significance of the data generated. Time values generated by a "black box" with no insight into the process cannot be used in optimum fashion to achieve time management purposes. For this reason, the manual computation of activity times is discussed in detail in the sections following. When these calculations are made by hand, they are normally performed directly on the network itself. When making the initial study, it is usual with manual computation that activity times be expressed in terms of <u>expired</u> working days. Commensurately, the start of the work being planned is customarily taken to be at time zero.

After a time duration has been estimated for each activity, some simple step-by-step computations are then performed. The purpose of these calculations is to determine (1) the overall project completion time and (2) the time brackets within which each activity must be accomplished if this completion time is to be met. The network calculations involve only additions and subtractions and can be made in different ways, although the data produced are comparable in all cases. The usual procedure is to calculate what are referred to as "activity times." When arrow notation is being used, so-called "event times" can also be used as a basis for network computations (see Section 6.10). Activity times play a fundamental role in project scheduling, and their determination is treated in this chapter. Event time computations are discussed in the next chapter.

The calculation of activity times involves the determination of four limiting times for each network activity. The "early start" (ES) or "earliest start" of an activity is the earliest time that the activity can possibly start allowing for the times required to complete the preceding activities. The "early finish" (EF) or "earliest finish" of an activity is the earliest possible time that it can be completed and is determined by adding that activity's duration to its early start time. The "late finish" (LF) or "latest finish" of an activity is the very latest that it can finish and allow the entire project to be completed by a designated time or date. The "late start" (LS) or "latest start" of an activity is the latest possible time that it can be started if the project target completion date is to be met and is obtained by subtracting the activity's duration from its latest finish time.

The computation of activity times can be performed manually or by computer. When calculations are made by hand, they are normally performed directly on the network itself. When the computer is used, activity times are shown on the computer generated network or in tabular form. When making manual computations it is normal to use project days, working in terms of expired working days. Commensurately, the start of the project is customarily taken to be time zero (the end of project day zero). When computations are done by computer, calendar dates are often used so that early and late starts are in the morning of the calendar day and early and late finishes are in the afternoon of the calendar day.

5.8 EARLY ACTIVITY TIMES

The highway bridge network shown in Figure 5.1 is used here to discuss the manual calculation of activity times directly on a precedence diagram. Precedence diagrams are exceptionally convenient for the manual calculation of activity times and afford an excellent basis for describing how such calculations are done. The computation of the early start and early finish times will be treated in this section. The determination of late activity times will be described subsequently. The early time computations proceed from project start to project finish, from left to right in Figure 5.1, this process being termed the "forward pass." The basic assumption for the computation of early activity times is that every activity will start as early as possible. That is to say, each activity will start just as soon as the last of its predecessors is finished.

The ES value of each activity is determined first, with the EF time then being obtained by adding the activity duration to the ES time. To assist the reader in understanding how the calculations proceed, small sections of Figure 5.1 will be used in Figure 5.2 to illustrate the numerical procedures. Reference to Figure 5.1 shows that activity 0 is the initial activity. Its earliest possible start is, therefore, zero elapsed time. As explained by the sample activity shown in Figure 5.1, the ES of each activity is entered in the upper left of its activity box. The value of zero is commensurately entered at the upper left of activity 0 in Figure 5.2(a). The EF of an activity is obtained by adding the activity duration to its ES value. Activity 0 has a duration of zero. Hence, the EF of activity 0 is its ES of zero added to its duration of zero or a value of zero. EF values are entered into the upper right of activity boxes, and Figure 5.2(a) shows the EF value of activity 0 to be zero. Activity 0 calculations are trivial, but the use of a single opening activity is customary with precedence diagrams.

Figure 5.1 shows that activities 10, 20, 30, 40, and 50 can all start after activity 0 has been completed. In going forward through the network, the earliest that these five activities can start is obviously controlled by the EF of the preceding activity. Since activity 0 has an EF equal to zero, then each of the following five activities can start as early as time zero. Figure 5.2(b) shows that activities 10, 20, 30, 40, and 50 all have ES values of zero, using activities 30 and 40 as typical examples. Zeros have been entered, therefore, in the upper left of activities 30 and 40. The EF of activity 30 is its ES of zero, plus its duration of 15 or a value of 15. Likewise, the EF value of activity 40 is zero, plus 3 or a value of 3.

Continuing into the network, Figure 5.1 and Figure 5.2(c) show that none of activities 80, 90, or 100 can start until activity 40 has been completed. Activity 40 is referred to as a "burst activity," this being an activity that is followed by two or more activities. The earliest that activity 40 can finish is at the end of the third day. Using

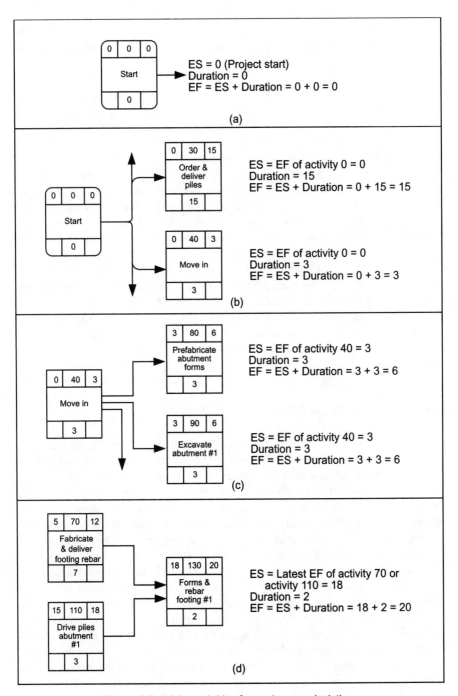

Figure 5.2   Highway bridge forward-pass calculations.

activities 80 and 90 as examples, it is seen that the earliest that these two activities can be started is after the expiration of three working days (or at a time of three). The EF of activity 80 will be its ES value of 3, plus its duration of 3 or a value of 6. In like fashion, activity 90 will also have an EF value of 6. These values have been entered into the activity boxes in Figure 5.2(c).

Figures 5.1 and 5.2(d) show that activity 130 cannot start until both activities 70 and 110 are finished. The EF values of activities 70 and 110 are 12 and 18, respectively. Because the ES of activity 130 depends on the completion of both activities, it follows that the finish of activity 110, not that of activity 70, actually controls the start of activity 130 and that activity 130 will have an early start of 18. Activity 130 is an example of a "merge activity," this being an activity whose start depends on the completion of two or more antecedent activities. The rule for this and other merge activities is that the earliest possible start time of such an activity is equal to the latest (or largest) of the EF values of the immediately preceding activities.

The forward-pass calculations consist of only repeated applications of the few simple rules just discussed. Working methodically in step-by-step fashion, the computations in Figure 5.1 proceed from activity to activity until the end of the network is reached. The reader is reminded that the figures and calculations shown in Figure 5.2 were for explanatory purposes only. The actual forward-pass calculations would have commenced with activity 0 in Figure 5.1 with the ES and EF values being entered onto the activity boxes as the calculations progressed. As is now obvious, the calculations are elementary and, with practice, can be performed rapidly. Even so, when several hundred activities are involved, the manual development of activity times can be tedious, time consuming, and subject to errors.

5.9 PROJECT DURATION

Reference to Figure 5.1 will disclose that the early finish time for the last work activity (400) is 64 elapsed working days. For the job logic established and the activity durations estimated, it will require 64 working days to reach the end of the project, provided that each activity is started as soon as possible (or at its ES time). What this says is that if a competent job of planning has been done, if activity durations have been accurately estimated, and if everything goes well in the field, project completion can be anticipated in 64 working days or about $7/5(64) \approx 90$ calendar days. In this regard, any labor holidays that occur on regular workdays must be added to the 90 calendar days just obtained. As will be seen later in Section 5.16, the construction period for the highway bridge will be during the months of June into September. During this time, the holidays of July 4 and Labor Day will occur. Therefore, the construction period will require approximately 92 calendar days.

The matter of contingency must again be considered at this point. Although allowances for lost time caused by inclement weather can be ascribed to the affected activities in some cases, an overall weather contingency is more appropriate in other cases (see Section 5.6). In addition, some provision must be made for general project delays caused by a variety of troubles, mischances, oversights, difficulties, and job casualties. Many contractors will, at this stage of the project scheduling, plan on a time overrun of 5 to 10 percent and add this onto the overall projected time requirement for the entire work. The percentage actually added must be based on a contractor's judgment and experience. In Figure 5.1, a contingency of six working days has been added to the diagram in the form of the final contingency activity 410. Adding an overall contingency of six working days gives a probable job duration of 70 working days. In the contractor's way of thinking, 70 working days represents a more realistic estimate of actual project duration than does the value of 64. If everything goes as planned, the job will probably be finished in about 64 working days. However, if the usual difficulties arise, the contractor has allowed for a 70 working-day construction period.

Whether the contractor chooses to add in a contingency allowance or not, now is the time to compare the computed project duration with any established project time requirement. Following along with the highway bridge and assuming that the contingency of six working days will be used, the figure of 7/5(70) plus 2 labor holidays = 100 calendar days is compared with the completion date established by the Example Project master schedule or by any time provision in the construction contract applicable to the highway bridge. If the highway bridge must be completed in 90 calendar days, then the contractor will have to consider ways in which to shorten its time duration. If a construction period of 100 calendar days is admissible, the contractor will feel reasonably confident that it will be able to meet this requirement and no action to shorten the work is required. How to go about decreasing a project's duration is the subject of Chapter 7.

The probable project duration of 100 calendar days or approximately 14 weeks is a valuable piece of information. For the first time the contractor has an estimate of overall project duration that it can rely on with considerable trust. The difference in confidence level between the value of 14 weeks that has just been obtained and the 15 weeks derived earlier from the bar chart in Figure 3.5 should be apparent.

5.10 LATE ACTIVITY TIMES

For purposes of discussion in the remainder of this chapter, it is assumed that a project duration of 100 calendar days or 70 working days for the highway bridge is acceptable. Unless there is some mitigating circumstance, there is no point in the contractor's attempting to rush the job and there certainly is nothing to be gained by deliberately allowing the work to drag along. The normal activity durations used as a scheduling basis represent efficient and economical operation. Shorter activity times will usually require expensive expediting actions. Longer activity times suggest a relaxed attitude and increased costs of production. Certainly, job overhead expense increases with duration of the construction period.

Having established that 70 working days is satisfactory, job calculations now "turn around" on this value and a second series of calculations is performed to find the late start (LS) and the late finish (LF) times for each activity. These calculations, called the "backward pass," start at the project end and proceed backward through the

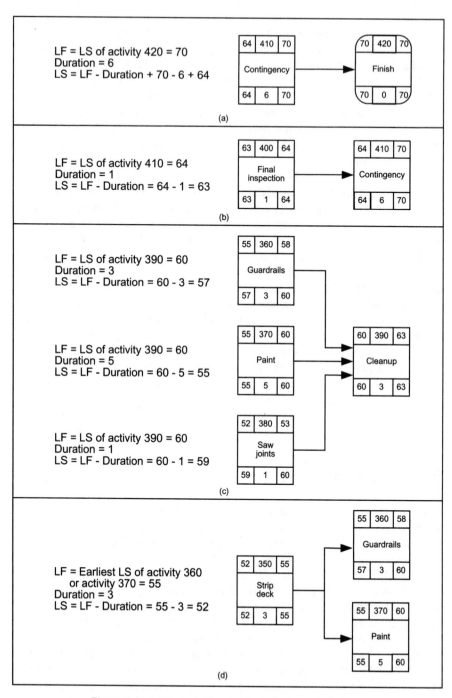

Figure 5.3 *Highway bridge, backward-pass calculations.*

network, going from right to left in Figure 5.1. The late activity times now to be computed are the latest times at which the several activities on the highway bridge can be started and finished with project completion still achievable in 70 working days. The supposition during the backward pass is that each activity finishes as late as possible without delaying project completion. The LF value of each activity is obtained first and is entered into the lower right portion of the activity box. The LS, in each case, is obtained by subtracting the activity duration from the LF value. The late start time is then shown at the lower left. Small sections of Figure 5.1 will be used in Figure 5.3 to help illustrate the numerical procedures.

The backward pass through Figure 5.1 is begun by giving activity 420 a LF time of 70. Figure 5.3(a) shows the value of 70 entered in the lower right of activity 420. The LS of an activity is obtained by subtracting the activity duration from its LF value. Activity 420 has a duration of 0. Hence, the LS of activity 420 is its LF of 70 minus its duration of 0 or a value of 70. This value of 70 has been entered at the lower left of activity 420 in Figure 5.3(a). Here again, this is a trivial calculation but it is customary to end precedence diagrams with a single closing activity. Continuing with Figure 5.3(a), it shows that activity 410 immediately precedes activity 420. In working backward through the network as shown in Figure 5.3(a), the latest that activity 410 can finish is obviously controlled by the LS of its succeeding activity 420. If activity 420 must start no later than day 70, then activity 410 must finish no later than that same day. Consequently, activity 410 has an LF time equal to the LS of the activity following (420) or a value of 70. With a duration of 6, it has an LS value of 70 minus 6, or 64. These values are shown on activity 410 in Figure 5.3(a).

Figures 5.1 and 5.3(b) show that activity 400 immediately precedes activity 410. The latest that activity 400 can finish is controlled by the LS of its succeeding activity 410. If activity 410 must start no later than day 64, then activity 400 must finish no later than that same day. Consequently, activity 400 has an LF time equal to the LS of the activity following (410) or a value of 64. With a duration of 1, it has an LS value of 64 minus 1, or 63. These values are shown on activity 400 in Figure 5.3(b).

Figures 5.1 and 5.3(c) disclose that activity 390 is preceded by three activities: 360, 370, and 380. The LF of each of the three antecedent activities is set equal to the LS of activity 390 or day 60. Subtracting the activity durations from their LF values yields the LS times. The LS times for activities 360, 370, and 380 are, correspondingly, equal to 57, 55 and 59. These values have been entered in Figure 5.3(c)

Some explanation is needed when the backward pass reaches a "burst activity," which is one that has more than one activity immediately following it. In Figure 5.1, activity 350 would be the first such activity reached during the backward pass, this activity being followed immediately by activities 360 and 370. To obtain the late finish of activity 350, the late starts of the immediately succeeding activities are noted. These are obtained from Figure 5.1 as 57 for activity 360 and 55 for activity 370 and are entered into Figure 5.3(d). Keeping in mind that activity 350 must be finished before either activity 360 or 370 can begin, it is obvious that activity 350 must be finished by no later than day 55. If it is finished any later than this, the entire project will be delayed by the same amount. The rule for this and other burst activities is that the LF value for such an activity is equal to the earliest (or smallest) of the LS times of the activities following.

The backward-pass computations proceed from activity to activity until the start of the project is reached. All that is involved are repetitions of the rules just discussed. In making actual network calculations, of course, the LF and LS values would be calculated directly on Figure 5.1 with the times being entered into the activity boxes as they are obtained. The reader is encouraged to verify all of the early and late activity times shown in Figure 5.1 as a test of how well the rules of computational procedure have been mastered.

5.11 TOTAL FLOAT

Examination of the activity times appearing in Figure 5.1 discloses that the early and late start times (also early and late finish times) are the same for certain activities and not for others. The significance of this is that there is leeway in the scheduling of some activities and none at all in the scheduling of others. This leeway is a measure of the time available for a given activity above and beyond its estimated duration. This extra time is called "float," two classifications of which are in general usage: total float and free float.

The total float of an activity is obtained by subtracting its ES time from its LS time. Subtracting the EF from the LF gives the same result. Once the activity times have been computed on a precedence diagram, values of total float are easily computed and may be noted on the network if desired. This has not been done on Figure 5.1, however, in an attempt to keep the figure as simple as possible. Referring to Figure 5.1, the total float for a given activity is found as the difference between the two times at the left of the activity box or between the two at the right. The same value is obtained in either case. An activity with zero total float has no spare time and is, therefore, one of the operations that controls project completion time. For this reason, activities with zero total float are called "critical activities." The second of the common float types, "free float," will be discussed in Section 5.13.

5.12 THE CRITICAL PATH

In a precedence diagram, a critical activity is quickly identified as one whose two start times at the left of the activity box are equal. Also equal are the two finish times at the right of the activity box. Inspection of the activities in Figure 5.1 discloses that there are 18 activities that have total float values of zero. Plotting these on the figure discloses that these 18 activities form a continuous path from project beginning to project end, this chain of critical activities being called the "critical path." The critical path is normally indicated on the diagram in some distinctive way such as with heavy lines which are used in Figure 5.1.

Inspection of the network diagram in Figure 5.1 shows that numerous paths exist between the start and end of the diagram. These paths do not represent alternate choices through the network. Rather, each of these paths must be traversed during the actual construction process. If the time durations of the activities forming a continuous path were to be added for each of the many possible routes through the network, a number of different totals would be obtained. The largest of these totals is the critical or minimum time for overall project completion. Each path must be traveled, so the longest of these paths determines the length of time necessary to complete *all* of the activities in accordance with the established project logic.

If the total times for all of the network paths in Figure 5.1 were to be obtained, it would be found that the longest path is the critical path already identified using zero total floats and that its total time duration is 70 days. Consequently, it is possible to locate the critical path of any network by merely determining the longest path. However, this is not usually a practical procedure. The critical path is normally found by means of zero total float values. It needs to be pointed out that the scheduler can be badly fooled if he attempts to prejudge which activities will be critical or to locate the critical path by inspection. Critical activities are not necessarily the most difficult or those that may seemingly be the most important job elements. Although there is only one critical path in Figure 5.1, more than one such path is always a possibility in network diagrams. One path can branch out into a number of paths, or several paths can combine into one. In any event, the critical path or paths must consist of an unbroken chain of activities from start to finish of the diagram. There must be at least one such critical path, and it cannot be intermittent. A break in the path indicates an error in the computations. On the highway bridge, 15 of the 40 activities (exclusive of start, contingency, and finish), or about 38 percent, are critical. This is considerably higher than is the case for most construction networks because of the small size of the highway bridge. In larger diagrams, critical activities generally constitute 20 percent or less of the total.

Any delay in a critical activity automatically lengthens the critical path. Because the length of the critical path determines project duration, any delay in the finish date of a critical activity, for whatever reason, automatically prolongs project completion by the same amount. As a consequence of this, identification of the critical activities is an important aspect of job scheduling because it pinpoints those job areas that must be closely monitored at all times if the project is to be kept on schedule.

5.13 FREE FLOAT

Free float is another category of spare time. The free float of an activity is found by subtracting its earliest finish time from the earliest start time of the activities directly following. To illustrate how free floats are computed, consider activity 90 in Figure 5.1 that shows the earliest finish time of activity 90 to be 6. The immediately following activities are 110 and 120, and the earliest start time of either is 6. The difference between the earliest finish date of activity 90 (day 6) and the earliest start time of the immediately following activities (day 6) is zero. Hence activity 90 has a free float of zero. Another example could be activity 270 that has an early finish time of 37. The earliest start date of the following activities, 360 and 370, is 55. Thus activity 270 has a free float value of 55–37 = 18 days. As an alternate statement of procedure, the free float of a given activity can be obtained by subtracting its EF (upper right) from the smallest of the ES values (upper left) of those activities immediately following.

The free float of an activity is the amount by which the completion of that activity can be deferred without delaying the early start of the following activities or affecting any other activity in the network. To illustrate, activity 270, which has a free float of 18 days, can have its completion delayed by 18 days because of late start, extended duration, or any combination thereof, without affecting any other network activity.

5.14 ACTIVITY TIME INFORMATION

Information concerning activity times and float values can be presented in three ways: on the network diagram itself, in tabular format, and in the form of bar charts. Using the precedence diagram of the highway bridge as a basis, the manual computation of activity times and floats has been presented in the preceding sections. When such computations are made, they are almost always performed directly on the network. This procedure is faster, more convenient, and conducive to greater accuracy.

To serve a variety of purposes, activity times and floats can be presented in table form as has been done in Figure 5.4. Such a table of values serves to collect together pertinent time information of the project into a form that is useful and convenient. In Figure 5.4, Column 2 lists the activity numbers as they appear in the precedence diagram of Figure 5.1 with the critical activities appearing in boldface type. The activity numbers listed in Column 3 are used in conjunction with Chapter 6 when arrow notation is discussed.

Activity time data presented in bar chart form are widely used for project time control purposes during the construction process. Project time information in the form of bar charts is discussed in Section 5.28.

5.15 FLOAT PATHS

All paths through the diagram in Figure 5.1, excepting the critical path, have summations of activity times less than 70 working days and are called "float paths." The longest float path, or the one with the least float, can be determined by referring to the total float values of Figure 5.4 and noting that one activity (360) has a total float value of only two days. This indicates that the next longest path through the network is $70 - 2 = 68$ working days. Reference to Figure 5.1 will show that this path is the same as the critical path except that activity 360 is substituted for activity 370. This path is said to have a float value of 2. The total float values of 3 in Figure 5.4 indicate that the next longest path through the network totals 67 days. As a matter of fact, there are two different paths through the network, each with a cumulative time total of 67 days or with float times of 3. Float paths can be located by linking activities with the same total floats together with critical activities as needed to form a continuous chain through the entire diagram.

When beginning a backward pass through a construction network, an established target duration for the project may be used for the "turn around" rather than the EF of the terminal activity. In such a case, it is possible for all paths through the network to be float paths. To illustrate, suppose that the backward pass through the highway bridge, Figure 5.1, had started with the LF of terminal activity 420 equal to 74 rather than 70. The effect of this on the values shown in Figure 5.4 would be to increase all of the late activity times (Columns 7 and 8) and all of the total float values by the constant amount of 4. In this case, there would be no activities with zero total floats. Nevertheless, there is still a critical path, this being the float path that has a minimum and constant total float value of 4. This path would be the same as the original critical path. The only difference is that each critical activity would now have a total float of 4 rather than zero. In a similar manner, if 67 had been used as the LF of terminal activity 420, all late activity times and total floats would be reduced by 3. In this case, the critical path would have a constant total float of −3 and would be the same critical path as that originally found.

5.16   EARLY START SCHEDULE

After the network calculations have been completed, the resulting activity times are used to prepare various forms of calendar-date schedules that are used for project time management. One of these is a schedule of activities based on their early start and finish times. This time schedule is called an "early start schedule" or a "normal schedule." The subject of preparing field operation schedules is discussed in Chapter 9. The treatment of early start schedules contained in this chapter is limited to introducing the general concept of project schedules and explaining how expired working days are converted to calendar dates.

Activity times obtained by manual calculations are expressed in terms of expired working days. For purposes of project monitoring and control it is necessary to convert these times to calendar dates on which each activity is expected to start and finish. This is easily done with the aid of a calendar on which the working days are numbered consecutively, starting with number 1 on the anticipated start date and skipping weekends and holidays (July 5 and September 6 in our case). Figure 5.5 is the conversion calendar for the highway bridge, assuming that the starting date is to be on Monday, June 14. Calendar dates appear in the upper left-hand corner of each box, and working days are circled.

The reader is reminded at this point that each major portion of the Example Project will be described and scheduled using its own unique planning network. When one is making manual time computations, each of the networks will begin at zero expired working days. It is at the current stage of conversion of expired working days to calendar dates that the time relationships among the various networks will be established. There will be a different Figure 5.5 for each network from which its unique calendar date time schedule is obtained.

When making up a job calendar, the true meaning of elapsed working days must be kept in mind. To illustrate, the early start of activity 180 in Figure 5.4 is 25. This means this activity can start after the expiration of 25 working days, so the starting date of activity 180 will be the morning of calendar date numbered 26. From Figure 5.5, working day 26 equates to the calendar date of July 20. There is no such adjustment for early finish dates.

Activity (Bold type denotes critical activities) (1)	Activity Number (2)	Activity Arrow (3)	Duration (Working Days) (4)	Earliest		Latest		Float	
				Start (ES) (5)	Finish (EF) (6)	Start (LS) (7)	Finish (LF) (8)	Total (TF) (9)	Free (FF) (10)
Start	**0**		**0**	**0**	**0**	**0**	**0**	**0**	**0**
Prepare & approve S/D abutment & deck rebar	**10**	**10-50**	**10**	**0**	**10**	**0**	**10**	**0**	**0**
Prepare & approve S/D footing rebar	20	10-60	5	0	5	9	14	9	0
Order & deliver piles	30	10-70	15	0	15	3	18	3	0
Move in	40	10-20	3	0	3	12	15	12	0
Prepare & approve S/D girders	50	10-40	10	0	10	8	18	8	0
Fabricate & deliver abutment & deck rebar	**60**	**50-160**	**15**	**10**	**25**	**10**	**25**	**0**	**0**
Fabricate & deliver footing rebar	70	60-100	7	5	12	14	21	9	6
Prefabricate abutment forms	80	20-160	3	3	6	22	25	19	19
Excavate abutment #1	90	20-30	3	3	6	15	18	12	0
Mobilize pile driving rig	100	20-70	2	3	5	16	18	13	10
Drive piles abutment #1	110	70-80	3	15	18	18	21	3	0
Excavate abutment #2	120	30-90	2	6	8	19	21	13	10
Forms & rebar footing #1	130	100-110	2	18	20	21	23	3	0
Drive piles abutment #2	140	90-140	3	18	21	21	24	3	0
Pour footing #1	150	110-120	1	20	21	23	24	3	0
Demobilize pile driving rig	160	140-160	1	21	22	24	25	3	3
Strip footing #1	170	120-130	1	21	22	24	25	3	0
Forms & rebar abutment #1	**180**	**160-180**	**4**	**25**	**29**	**25**	**29**	**0**	**0**
Forms & rebar footing #2	190	150-170	2	22	24	30	32	8	0

5.16 EARLY START SCHEDULE **91**

Activity	ID	Activity span	Dur	ES	EF	LS	LF	TF	FF
Pour abutment #1	**200**	**180-200**	**2**	**29**	**31**	**29**	**31**	**0**	**0**
Pour footing #2	210	170-190	1	24	25	32	33	8	0
Strip & cure abutment #1	**220**	**200-210**	**3**	**31**	**34**	**31**	**34**	**0**	**0**
Strip footing #2	230	190-220	1	25	26	33	34	8	8
Forms & rebar abutment #2	**240**	**220-230**	**4**	**34**	**38**	**34**	**38**	**0**	**0**
Pour abutment #2	**250**	**230-240**	**2**	**38**	**40**	**38**	**40**	**0**	**0**
Fabricate & deliver girders	260	40-260	25	10	35	18	43	8	8
Rub concrete abutment #1	270	210-300	3	34	37	52	55	18	18
Backfill abutment #1	280	210-260	3	34	37	40	43	6	6
Strip & cure abutment #2	**290**	**240-250**	**3**	**40**	**43**	**40**	**43**	**0**	**0**
Rub concrete abutment #2	300	250-300	3	43	46	52	55	9	9
Backfill abutment #2	310	250-280	3	43	46	46	49	3	3
Set girders	**320**	**260-270**	**2**	**43**	**45**	**43**	**45**	**0**	**0**
Deck forms & rebar	**330**	**270-280**	**4**	**45**	**49**	**45**	**49**	**0**	**0**
Pour & cure deck	**340**	**280-290**	**3**	**49**	**52**	**49**	**52**	**0**	**0**
Strip deck	**350**	**290-300**	**3**	**52**	**55**	**52**	**55**	**0**	**0**
Guardrails	360	300-310	3	55	58	57	60	2	2
Paint	**370**	**300-320**	**5**	**55**	**60**	**55**	**60**	**0**	**0**
Saw joints	380	290-320	1	52	53	59	60	7	7
Cleanup	**390**	**320-330**	**3**	**60**	**63**	**60**	**63**	**0**	**0**
Final inspection	**400**	**330-340**	**1**	**63**	**64**	**63**	**64**	**0**	**0**
Contingency	**410**	**340-350**	**6**	**64**	**70**	**64**	**70**	**0**	**0**
Finish	**420**		**0**	**70**	**70**	**70**	**70**	**0**	**0**

Figure 5.4 Highway bridge, activity times.

June

Su	M	Tu	W	Th	F	Sa
		1	2	3	4	5
6	7	8	9	10	11	12
13	14 (1)	15 (2)	16 (3)	17 (4)	18 (5)	19
20	21 (6)	22 (7)	23 (8)	24 (9)	25 (10)	26
27	28 (11)	29 (12)	30 (13)			

July

Su	M	Tu	W	Th	F	Sa
				1 (14)	2 (15)	3
4	5	6 (16)	7 (17)	8 (18)	9 (19)	10
11	12 (20)	13 (21)	14 (22)	15 (23)	16 (24)	17
18	19 (25)	20 (26)	21 (27)	22 (28)	23 (29)	24
25	26 (30)	27 (31)	28 (32)	29 (33)	30 (34)	31

August

Su	M	Tu	W	Th	F	Sa
1	2 (35)	3 (36)	4 (37)	5 (38)	6 (39)	7
8	9 (40)	10 (41)	11 (42)	12 (43)	13 (44)	14
15	16 (45)	17 (46)	18 (47)	19 (48)	20 (49)	21
22	23 (50)	24 (51)	25 (52)	26 (53)	27 (54)	28
29	30 (55)	31 (56)				

September

Su	M	Tu	W	Th	F	Sa
			1 (57)	2 (58)	3 (59)	4
5	6	7 (60)	8 (61)	9 (62)	10 (63)	11
12	13 (64)	14 (65)	15 (66)	16 (67)	17 (68)	18
19	20 (69)	21 (70)	22	23	24	25
26	27	28	29	30		

Figure 5.5 *Highway bridge, conversion calendar.*

In the case of activity 180, its early finish time is 29, which indicates that it is finished by the end of the twenty-ninth working day. Hence from Figure 5.5, the early finish date of that activity will be the afternoon of July 23.

It has been mentioned previously that calendar date information concerning network activities can be presented in different ways, depending upon the use to be made of the data. Figure 5.6 (between pages 78 and 79) shows one way in which computers can enter the early start and finish dates of the individual activities on the precedence diagram for the highway bridge.

5.17 TABULAR TIME SCHEDULES

When the computed activity times are converted to calendar dates, this schedule information is often presented in tabular format. The table in Figure 5.7 is the early

PROJECT CALENDAR				
Activity (Bold denotes critical activities)	Activity Number	Duration (Working Days)	Scheduled Starting Date A.M.	Scheduled Completion Date P.M.
Prepare & approve S/D, abutment & deck rebar	**10**	**10**	**June-14**	**June-25**
Prepare & approve S/D, footing rebar	20	5	June-14	June-18
Order & deliver piles	30	15	June-14	July-2
Move in	40	3	June-14	June-16
Prepare & approve S/D, girders	50	10	June-14	June-25
Prefabricate abutment forms	80	3	June-17	June-21
Excavate abutment #1	90	3	June-17	June-21
Mobilize pile driving rig	100	2	June-17	June-18
Fabricate & deliver footing rebar	70	7	June-21	June-29
Excavate abutment #2	120	2	June-22	June-23
Fabricate & deliver abutment & deck rebar	**60**	**15**	**June-28**	**July-19**
Fabricate & deliver girders	260	25	June-28	August-2
Drive piles, abutment #1	110	3	July-6	July-8
Forms & rebar, footing #1	130	2	July-9	July-12
Drive piles, abutment #2	140	3	July-9	July-13
Pour footing #1	150	1	July-13	July-13
Demobilize pile driving rig	160	1	July-14	July-14
Strip footing #1	170	1	July-14	July-14
Forms & rebar, footing #2	190	2	July-15	July-16
Pour footing #2	210	1	July-19	July-19
Forms & rebar, abutment #1	**180**	**4**	**July-20**	**July-23**
Strip footing #2	230	1	December-21	July-20
Pour abutment #1	**200**	**2**	**July-26**	**July-27**
Strip & cure, abutment #1	**220**	**3**	**July-28**	**July-30**
Forms & rebar, abutment #2	**240**	**4**	**August-2**	**August-5**
Rub concrete, abutment #1	270	3	August-2	August-4
Backfill abutment #1	280	3	August-2	August-14
Pour abutment #2	**250**	**2**	**August-6**	**August-9**
Strip & cure, abutment #2	**290**	**3**	**August-10**	**August-12**
Rub concrete, abutment #2	300	3	August-13	August-17
Backfill abutment #2	310	3	August-13	August-17
Set girders	**320**	**2**	**August-13**	**August-16**
Deck forms & rebar	**330**	**4**	**August-17**	**August-20**
Pour & cure deck	**340**	**3**	**August-23**	**August-25**
Strip deck	**350**	**3**	**August-26**	**August-30**
Saw joints	380	1	August-26	August-26
Guardrails	360	3	August-31	September-2
Paint	**370**	**5**	**August-31**	**September-7**
Cleanup	**390**	**3**	**September-8**	**September-10**
Final inspection	**400**	**1**	**September-13**	**September-13**
Contingency	**410**	**6**	**September-14**	**September-21**

Figure 5.7 Highway bridge, early start schedule.

start schedule for the highway bridge with the activities being listed in order of their starting dates. These calendar dates are often referred to as "scheduled" or "expected" dates. Not only is this sort of operational schedule useful to the contractor, but it can also be used to satisfy the usual contract requirement of providing the owner and architect–engineer with a projected timetable of construction operations.

Whether the contractor prepares activity schedule data in the form of network diagram information or activity time tables depends upon the use for which the information is intended. To be noted is the fact that while tabular reports provide activity numbers, descriptions, schedule dates, and float information, they do not reflect project logic. Tabular reports, like bar charts, communicate only basic information concerning individual activities. They do not communicate the sequence of activities and, therefore, are not diagnostic tools. Only network diagrams have this capability. What this says is that the form of time control information provided to a member of the project management team must be selected to meet the demands and responsibilities of the position.

5.18 ACTIVITIES AND CALENDAR DATES

Until elapsed working days have been converted to calendar dates, there has been no accurate way to associate activities with calendar times. In the case of short duration work, such as the highway bridge, this has been no problem because it was recognized from the beginning that the work would be done during the summer months. However, on projects requiring many months or years, it is important to associate general classes of work with the seasons of the year during which the work will be performed. The general time schedule developed during project cost estimating has provided guidance in this regard. Nevertheless, this preliminary construction schedule is approximate, at best, and may have been altered during project planning. Consequently, the first version of the working job calendar must be examined with the objective of comparing activities with the weather to be expected during their accomplishment. This perusal might well disclose some activities that should be expedited or delayed to avoid cold or wet weather, spring runoff, or other seasonal hazards. It may reveal a need for cold weather operations hitherto unsuspected. It can be a good guide for the final inclusion of weather contingency allowances.

5.19 SORTS

To serve a variety of different purposes, network activities can be listed in several different manners. The ways that activities are grouped together or the orders in which they are listed are called "sorts." Sorting provides emphasis on different criteria and makes different forms of information easier and quicker to find. Computers have the capability of sorting large numbers of activities quickly and accurately. By sorting activities on different bases, different forms of useful information become available to project management personnel. The following lists a number of examples.

1. Activity number sort: In this case, the activities are listed in ascending order of their assigned numbers as was the case in Figure 5.4. This sort eases going back and forth from the network diagram to the activity data. With the activity number from the diagram, the schedule dates and float values are quickly found from the corresponding activity number in this sort listing.

2. Early start sort: Activities listed in the calendar sequence of their earliest possible start times is an optimistic schedule and is commonly used by project field personnel. Figure 5.7 is an example of this type of sort. Although it is unlikely that all activities will be started by their early start dates, such a listing does serve as a daily reminder of the necessity of meeting these dates in the case of critical activities and of the fact that little time slippage can be tolerated with low-float activities. It also keeps field supervisors aware that float is being consumed by those activities whose beginnings are delayed beyond their early start dates. Such schedules are also desirable for vendors and subcontractors who are responsible for providing shop drawings, samples, and other submittal information. Design professionals should also work to an early start schedule so that such submittals will receive their timely attention.

3. Late start sort: The late start of an activity is the time by which it must be started if the project is not to be delayed. Failure of an activity to start by its late start date is the first indication that the project can be in time trouble.

4. Late finish sort: A late finish sort, one where the activities are listed in the order of their latest allowable finish dates, is a convenient monitoring device. If an activity has not been finished by its late finish date, the project is automatically behind schedule according to the established action plan.

5. Total float sort: A listing of activities in order of their criticality, that is, in ascending order of total float values, can be very helpful to job management in pinpointing those areas whose timely completions are of top priority. Figure 5.8, which is a partial listing of the activities of the highway bridge, is seen to bring into sharp focus the identity of those activities which require the closest attention insofar as timely completion is concerned. Those activities appearing at the top of such a list can be given the special attention they deserve. If an activity begins to slip timewise as the project progresses, its float will decrease and it will rise up the list on this type of report.

6. Project responsibility: Each network activity can be assigned the name of the organization or person responsible for its timely completion. Some activities will be assigned to the general contractor while others are the responsibility of a named subcontractor, the owner, or the design professional.

7. Combined sorts: A valuable aspect of sorting lies in combining sorting criteria. For example, sorts can be produced that will provide a specific job supervisor with a named contractor with a listing of those activities for which he is responsible. These activities can be ordered by early start and be limited to those activities that are scheduled to begin within the next 30 days. With this kind of sorting capability, it is possible to provide specific information to each of the people and organizations who share the responsibility for timely project completion.

The preceding discussion is intended to give the reader insight into the types of sorts commonly used and the project management uses to which they are put. It is possible to list many types of project information for use by different members of the management team.

Activity (Bold denotes critical activities)	Activity Number	Duration (Working Days)	Expected Start Date A.M.	Total Float
Prepare & approve S/D, abutment & deck rebar	**10**	**10**	**June-14**	**0**
Fabricate & deliver abutment & deck rebar	**60**	**15**	**June-28**	**0**
Forms & rebar, abutment #1	**180**	**4**	**July-20**	**0**
Pour abutment #1	**200**	**2**	**July-26**	**0**
Strip & cure, abutment #1	**220**	**3**	**July-28**	**0**
Forms & rebar, abutment #2	**240**	**4**	**August-2**	**0**
Pour abutment #2	**250**	**2**	**August-6**	**0**
Strip & cure, abutment #2	**290**	**3**	**August-10**	**0**
Set girders	**320**	**2**	**August-13**	**0**
Deck forms & rebar	**330**	**4**	**August-17**	**0**
Pour & cure deck	**340**	**3**	**August-23**	**0**
Strip deck	**350**	**3**	**August-26**	**0**
Painting	**370**	**5**	**August-31**	**0**
Cleanup	**390**	**3**	**September-8**	**0**
Final inspection	**400**	**1**	**September-13**	**0**
Guardrails	360	3	August-31	2
Order & deliver piles	30	15	June-14	3
Drive piles, abutment #1	110	3	July-6	3
Forms & rebar, footing #1	130	2	July-9	3
Drive piles, abutment #2	140	3	July-14	3
Pour footing #1	150	1	July-13	3
Demobilize pile driving rig	160	1	July-14	3
Strip footing #1	170	1	July-14	3
Backfill abutment #2	310	3	August-13	3

Figure 5.8 Highway bridge, total float sort.

5.20 LAGS BETWEEN ACTIVITIES

All previous discussions of project planning and scheduling have been on the basis of two important assumptions. One is that an activity cannot start until *all* the *immediately* preceding activities have been completed. The other is that once all antecedent activities have been finished, the following activity can start immediately thereafter. Although these assumptions are more or less true for most network relationships, there are instances where they are not. Consequently, a more flexible notation convention has been developed that can be used to show these more complex activity relationships. Such a system involves the use of lags between activities.

Figure 5.9 presents several examples of the use of lags. Unfortunately, nomenclature in this area is not standardized and "lead times" are often used to describe exactly the same precedence relationships. This is merely a matter of which activity is taken as the reference. A successor "lags" a predecessor, but a predecessor "leads" a successor. In this text, only lag times are used. Lag time can be designated on a dependency line with a positive, negative, or zero value. If no time is designated, it is assumed to be zero. In effect, a negative lag time is a lead time.

It is not intended that the activities and activity times shown in Figure 5.9 relate or pertain to the highway bridge. These are just general examples designed to illustrate each particular time relationship. The activity times shown in Figure 5.9

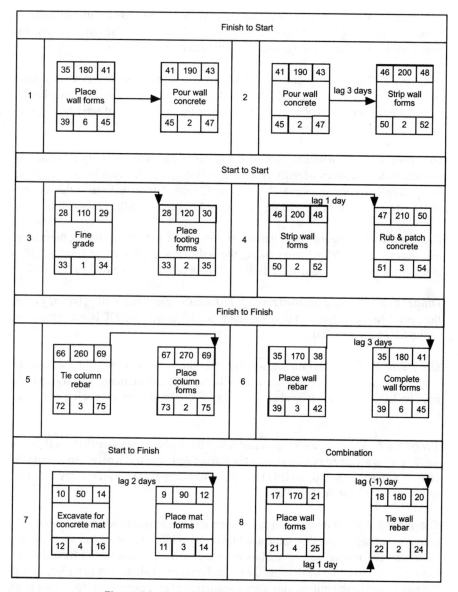

Figure 5.9 *Lag relationships, precedence notation.*

illustrate how the forward-pass and the backward-pass computations are performed between the two activity boxes affected.

Diagram 1. This figure shows that the succeeding activity 190 can start no earlier than the completion of activity 180. Here, there is no value of lag time indicated which means a value of zero or that there is no lag between the finish of activity 180 and the beginning of activity 190. This figure indicates that the wall concrete can be poured immediately after the wall forms have been completed with no delay in between. This is the usual form of activity dependency and is the only one that has been used in the network diagrams presented thus far in this text.

Diagram 2. This figure indicates that activity 200 cannot start until three days after activity 190 has been completed. This condition is reflected in the time calculations by making the early start of activity 200 equal to the early finish of activity 190, plus the time delay of three days or a value of 46.

Diagram 3. The dependency arrow in this figure shows that activity 120 can start no earlier than the start of activity 110. The fact that there is no lag value shown (hence a value of zero) indicates that activity 120 can start immediately after the start of activity 110. In this case, the ES of activity 120 is equal to the ES of activity 110.

Diagram 4. This figure shows that after one day of stripping wall forms, the cement masons can start patching and rubbing the wall surfaces. The ES of activity 210 is made equal to the ES value of activity 200, plus one day of lag or a value of 47.

Diagram 5. Here, the finish of activity 270 occurs immediately after, but only after, the completion of activity 260. In this case, the EF of activity 270 is made the same as the EF of activity 260.

Diagram 6. This diagram shows that the finish of activity 180 follows the completion of activity 170 by three days. In the time calculations, activity 180 will have an EF equal to the EF value of activity 170, plus the three days lag or a value of 41.

Diagram 7. The start-to-finish dependency shown here indicates that the finish of activity 90 is achieved two days after the start of activity 50. Two days following the start of activity 50 will be its ES (10 expired working days) plus the delay of 2 days or a value of 12 for the EF of activity 90.

Diagram 8. The combination of lags shown indicates the following conditions. The start-to-start lag of one day indicates that tieing reinforcing steel can start one day after the wall forming has begun The finish-to-finish dependency of −1 (note that this lag is minus and could be described as a lead time of +1) indicates that placing wall forms cannot finish until one day after completion of the reinforcing steel placement.

Any construction network involves many simplifications and approximations of reality. Certainly among these is the usual assumption that an instantaneous transition occurs between a completed activity and those that immediately follow. It is undoubtedly true that most sequential transitions between successive activities on actual construction jobs involve either some overlapping of one another or some delay between the finish of one and start of the next. Commensurately, absolute precision in making up project networks would necessitate the widespread usage of lag relationships. This would substantially complicate the planning and scheduling process with little gain in management efficacy. For this reason, lag times are not extensively used except where the time effects are substantial or for special construction types. The use of lags is especially convenient when working with long strings of simultaneous and repetitive operations such as the pipeline relocation.

5.21 PIPELINE SCHEDULING COMPUTATIONS

Network computations for projects that involve repetitive operations proceed in precisely the same manner as for any other project. Figure 5.10, which depicts the same job logic as Figure 4.5, illustrates the determination of activity times for the pipeline relocation discussed in Section 4.13. In actual practice, the contractor would undoubtedly have the crews and equipment in better balance than that indicated by the activity times used in Figure 5.10 (between pages 104 and 105). In other words, the times to accomplish a mile of location and clearing, a mile of excavation, and a mile of the other operations would be about the same. This obviously would help to prevent the undesirable situation of having one operation unduly limit another with attendant wasted time and loss of operational efficiency. The activity times used in Figure 5.10 were selected for purposes of illustrating the generality of the procedure. The bold print activities in Figure 5.10 constitute the critical path, located by those activities with zero total float.

5.22 PIPELINE SUMMARY DIAGRAM

Summary diagrams for repetitive operation projects are of particular importance. For example, suppose that the pipeline in Figure 5.10 were 20 miles in length rather than 5. A detailed figure such as Figure 5.10 for the entire project length would be impossibly large. Figure 5.11 is a condensation of Figure 5.10 obtained by using lag notation. Figure 5.11 shows all 5 miles of each individual operation as a single activity box and is of a type that lends itself well to long strings of repeated operations. For example, activity 10 in this figure represents the locating and clearing of the entire 5 miles of right-of-way and has a duration of five days. In a similar manner, activity 20 represents all the excavation and has a total time duration of 20 days. The other four job operations are shown in a similar manner. As explained previously, excavation and string pipe can proceed simultaneously with one another following the location and clearing.

To discuss the activity time computations with regard to Figure 5.11, reference is made to activity 40, which may be considered as typical.

Computation of ES. The ES of activity 40 is computed twice. One possible value is the ES of activity 20 added to the delay of 4 giving a value of 5. The other is the ES of activity 30 added to the delay of 2 or a value of 3. Since activity 40 cannot start until four days after activity 20 has started and two days after activity 30 has started, it is most severely restrained by activity 20 and, therefore, has an ES of 5.

Computation of EF. The EF of activity 40 is computed three times. First, the addition of the EF of activity 20 to the delay of 5 gives a value of 26. Second, adding

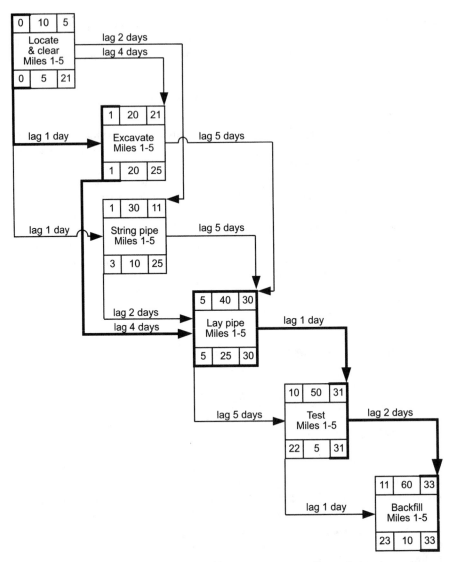

Figure 5.11 Pipeline relocation, summary precedence diagram.

the delay of 5 to the EF of activity 30 yields 16. Third, adding the ES of activity 40 to its duration of 25 gives 30. The largest value (30) is the EF value.

Computation of LF. The LF of activity 40 has only one possible value. This is computed by subtracting the lag of one day from the LF of activity 50, giving a value of 30.

Computation of LS. The LS of activity 40 has two possible values. First, the duration of activity 40 is 25. Subtracting this from its EF value of 30 yields 5. Second, the LS of activity 50 is 22. Subtracting the five-day lag from this gives 17. The LS of activity 40 is the smaller of 5 and 17 or a value of 5.

The use of such a summary diagram can sacrifice much of the internal logic. Reference is made to the location of the critical path in Figure 5.11, which does not reveal the same level of detail as does Figure 5.10. To illustrate this point, it is to be noted in Figure 5.11 that summary activities 10 and 20 have equal values of early start (ES) and late start (LS). However, there are different values of early finish (EF) and late finish (LF). These values indicate that part, but not all, of the Locate and clear sequence (activity 10) and of the Excavate sequence (activity 20) are critical. In themselves, these summary activities do not indicate how much of each sequence is critical. The heavy black line in Figure 5.11 merely shows that parts of these two work sequences are critical. Further study will disclose that only the first day of activity 10 and the first day of activity 20 are critical. The computed early and late times for activity 40 indicate that the entire Lay pipe sequence is critical, which is so indicated by a heavy black line around the entire activity box.

5.23 INTERFACE COMPUTATIONS

In Section 4.14, a pipeline crossing structure network was discussed that interfaces with the pipeline relocation network. The network for the pipeline crossing structure, which is located at the end of mile three of the pipeline relocation right-of-way, is shown in Figure 5.12. The interfacing activities between these two networks are activity 110 (Figure 5.12) and activity 170 (Figure 5.10). It has been decided that the crossing structure must be ready to accept pipe by the time the pipe-laying operation of the main pipeline reaches it. The interrelationship between the two interfacing activities is established in the following manner. Reference to Figure 5.10 shows that activity 170 has an EF = LF = 20 which means that the third mile of the pipeline will be laid and in place by elapsed working day 20. This simply says that the crossing structure must be able to accept pipe (activity 110 in Figure 5.12) by no later than 20 elapsed working days. Thus, the LS for activity 110 in Figure 5.12 is set equal to 20. When this is done, the time schedules for both the pipeline and crossing structure are meshed together and are on the same basis.

The LS value of 20 has been entered for activity 110 in Figure 5.12 and the forward pass completed from that point. A backward pass through the crossing network is now performed, showing that the LS value of the first activity (10) is 3. This signifies that the construction of the crossing structure must be started no later than the beginning of the fourth day (time = 3) if the crossing is to be ready for pipe when it is needed. This follows from the fact that the same time scale is being used for both the pipeline relocation and the crossing structure, that is, zero time is the same for both. Making a forward pass up to activity 110 using an ES of activity 10 in Figure 5.12 equal to zero yields the ES and EF times shown.

The total float values for each activity have also been determined and are shown in Figure 5.12. As can be seen, there is a continuous path through the network from activity 10 to activity 110 with a constant total float of 3. This float path is the critical path for that project. The significance of the float value of 3 is that if the crossing

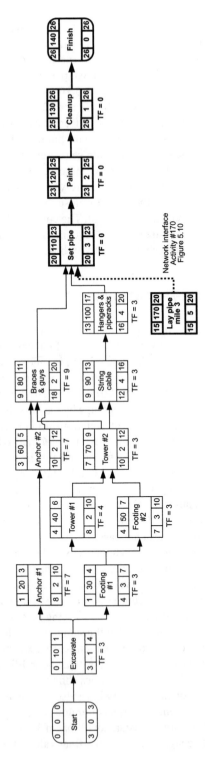

Figure 5.12 Pipeline crossing structure, precedence diagram time computations.

structure is started at the same time as the main pipeline (time = 0), the longest path through the network has three days of spare time associated with it. If everything goes reasonably smoothly, the crossing structure should be ready for pipe in 17 working days or three days ahead of the arrival of the pipeline itself. If troubles develop on the crossing structure, a general time contingency of three days has been provided.

If the EF of activity 170 in Figure 5.10 had a value of 16 rather than 20, all the LS and LF values in Figure 5.12 would be four less than those shown. The LS of activity 10 would have been –1, indicating that the crossing structure would have to be started one day before the main pipeline if the crossing is to be ready on time.

5.24   HAMMOCK ACTIVITY

A useful concept with regard to drawing project diagrams is the "hammock activity," one that extends from one activity to another, but which has no estimated time duration of its own. It is an activity in the usual sense in that it is time-consuming and requires resources, but its duration is controlled, not by its own nature, but by the two activities between which it spans. Its ES and LS times are determined by the activity where it begins and its EF and LF times are dictated by the activity at its conclusion. Common examples of hammock activities are dewatering and haul road maintenance, construction operations whose time spans are not self-imposed, but are dictated by other job factors.

As a specific illustration, suppose that construction of the highway bridge were to involve diversion of the stream during the construction and backfilling of the abutments. Reference to Figure 5.1 shows that the stream must be diverted prior to the start of activity 90, "Excavate abutment #1," and can be removed only after the completion of activity 310, "Backfill abutment #2." Figure 5.13 illustrates how hammock activity 35, "Maintain stream diversion," would span the time interval between activity 25, "Divert stream," and activity 275, "Remove stream diversion." Activity 35 has no intrinsic time requirement of its own, this being determined

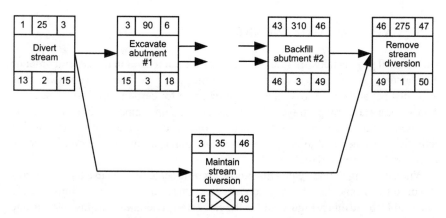

Figure 5.13   *Highway bridge, hammock activity.*

entirely by the finish of activity 25 and the start of activity 275. According to the present plan and schedule, the duration of activity 35 can vary anywhere from 34 to 43 working days, these values being obtained from the start and finish times of this activity in Figure 5.13.

5.25   MILESTONES

Milestones are points in time that have been identified as being important intermediate reference points during the accomplishment of the work. Milestone events can include dates imposed by the owner for the finishing of certain tasks as well as target dates set by the contractor for the completion of certain segments of the work. A milestone is usually the scheduled date for either the start or the completion of some difficult or important aspect of the project. On large projects, contractors frequently establish a series of milestones extending throughout the project and use these as reference points for project monitoring. The usage of milestones for job monitoring is discussed in Section 9.3.

Events, or points in time, do not appear as such on the customary precedence diagram. Nevertheless, milestones can be shown on a project network if this should be desired by project management. The usual convention is to show an event as a zero time duration box at the appropriate location within the diagram. If some distinctive geometric figure is preferred, circles, ovals, triangles, or other shapes can be used. Any information pertaining to a milestone and considered to be useful may be entered. To illustrate how a milestone can be indicated on a precedence network, suppose that a target date for the highway bridge job is when the abutments have been completed and backfilled so that water flow down the ravine being spanned by the bridge is no longer a potential problem. Reference to Figure 5.1 shows that this point in time is reached with the completion of activity 310. A symbolic convention commonly used for depicting milestones on precedence diagrams is shown in Figure 5.1 and 5.6. It is to be noted that in a network which numbers the activities by 10's, milestones can be depicted by numbers ending in 5's.

5.26   TIME-SCALED NETWORKS

The original project network is arranged to show its activities in the general order of their accomplishment, but is not plotted to a time scale. A time-scaled network has some distinct advantages over the regular diagram for certain applications because it provides a graphical portrayal of the time relationships among activities as well as their sequential order. Such a plot enables one to determine immediately which activities are scheduled to be in process at any point in time and to detect quickly where problem areas exist.

When drawing a time-scaled diagram, two time scales can be used: one in terms of working days and the other as calendar dates. One scale is, of course, immediately convertible to the other. Figure 5.14 is the early start schedule of the highway bridge plotted to a working day scale with the calendar months also indicated. This is the same network diagram as Figure 5.1, and the two portray the same logic and convey the same scheduling information.

Figure 5.14 is obtained by plotting, for each activity, its ES and EF values. Anchored at its left end by its ES time, each activity is visualized as being stretched to the right so that its horizontal length is equal to its estimated time duration. Commensurately, each activity is shown as a one dimensional line rather than as a two-dimensional box. Vertical solid lines indicate sequential dependence of one activity on another. When an activity has an early finish time that precedes the earliest start of activities following, the time interval between the two is, by definition, the free float of the activity. Free floats are shown as horizontal dashed lines in Figure 5.14, and a time-scaled plot of this type automatically yields to scale the free float of each activity. When an activity has no free float, no dashed extension to the right of that activity appears. The horizontal dashed lines also represent total float for groups or strings of activities, a topic that is discussed in Section 5.27.

The circles shown in Figure 5.14 are entered to clarify where one activity ends and another begins as well as to indicate sequential dependencies. The numerical values in the various circles are the ES and EF times in terms of expired working days for the respective activities.

Time-scaled diagrams are very convenient devices for checking daily project needs of labor and equipment and for the advance detection of conflicting demands among activities for the same resource. Chapter 8 discusses this subject in detail. A network plotted to a time scale is also useful for project financial management applications and for the monitoring of field progress. Summary diagrams for management, owners, and architect-engineers are frequently drawn to a time scale to enhance the diagram's ease of comprehension and ready application to job checking and evaluation. Milestones can be indicated on time-scaled networks using any desired distinctive symbol. By way of example, the triangle in Figure 5.14 at the finish of activity 310 indicates the previously discussed point in time when the abutments have been completed and backfilled.

5.27 NATURE AND SIGNIFICANCE OF FLOATS

A project network plotted to a time scale illustrates and clarifies the nature of free and total floats. In Figure 5.14, the critical path can be visualized as a rigid and unyielding frame extending through the diagram. The horizontal dashed lines (float) can be regarded as elastic connections between activities that can be shortened or elongated. The vertical solid lines represent dependencies among activities. Under circumstances to be discussed in later chapters, activities can be rescheduled provided that the correct dependency relationships are maintained and if there is sufficient float present to accommodate the change.

To illustrate the nature of free float, consider activity 310 in Figure 5.14. This activity has a duration of three days and a free float of three days. Figure 5.14 clearly shows that a total of six working days is available to accomplish this activity. Within its two time boundaries (day 43 and day 49), this activity can be delayed in starting, have its duration increased, or a combination of the two without disturbing any other activity. For all practical purposes, activity 310 can be treated as an abacus bead three units in length that can be moved back and forth on a wire six units long. The free float of an activity, therefore, is extra time associated with the activity that can be used or consumed without affecting the early start time of any succeeding activity.

The nature of total float can also be understood from a study of Figure 5.14. For example, activity 310 also has a total float of three days. If the completion of this activity were to be delayed by three days, it would become a critical activity and a new loop would materialize on the critical path. It follows from this demonstration that for a given activity, when free float is used, total float is also used in the same amount.

In most cases, however, the nature of total float is considerably more involved than the example just given. The problem with total float is that it is usually shared by strings or aggregations of activities. To maintain the necessary logic dependencies when using total float, one must often move whole groups of activities as a unit. The time-scaled plot is a convenient basis for doing this, but the process can become involved at times.

To illustrate a more complex case, reference is made to Figure 5.14 and activities 130, 150, and 170, each of which has a total float of three days. These three activities, as a group, have a combined flexibility of three days. If three days are lost in achieving completion of any of the activities of this string, the three days' flexibility is gone, the three days' total float of *all three* of the activities is consumed, and a new branch on the critical path is formed through activities 30–110–130–150–170. Reference to Figure 5.1 shows that activity 190 must follow activity 170. Consequently, when activity 170 moves three days to the right, Figure 5.14 reveals that activities 190, 210, and 230 do likewise. Thus all floats of these three activities are reduced from 8 to 5, and the free float of activity 140 is increased by 3 to a value of 4.

This discussion shows that the total float of an activity is the length of time that the early finish of the activity can be delayed and not adversely affect project completion. If all the total float of a given activity is consumed, accidentally or by design, the activity becomes critical and a new critical path, or branch thereon, is created in the network. In this new critical path, all the activities prior to the given activity must be completed by their EF times and all activities following will begin at their LS times.

5.28 BAR CHARTS

Bar charts, briefly discussed in Section 3.9, present the project schedule plotted to a horizontal time scale. The bar chart has been a traditional management device for planning and scheduling construction projects. However, bar charts have well recognized and serious shortcomings when used for the original development of project management information. For one, the interdependencies among activities are difficult to show and are often not reflected in the data generated. Additionally, the bar chart in itself does not provide a basis for ascertaining which activities are critical and which are "floaters." Consequently, each activity receives the same consideration with no indication of where management attention should be focused. The bar chart is not an adequate planning and scheduling tool because it does not portray a detailed, integrated, and complete plan of operations. Bar charts are completely ineffective for project shortening, resource management, and most of the other project management methods yet to be discussed.

However, the unsurpassed visual clarity of the bar chart makes it a very valuable medium for displaying job schedule information. It is immediately intelligible to people who have no knowledge of CPM or network diagrams. It affords an easy and convenient way to monitor job progress, schedule equipment and crews, and record project advancement. For these reasons, bar charts continue to be widely used in the construction industry. The use of bar charts as project time management devices is discussed in Chapter 9. While CPM networks are planning and diagnostic tools, bar charts are visual display devices.

Fortunately, it is possible to prepare bar charts on a more rational basis, avoiding their intrinsic weaknesses and incorporating the strengths and advantages of network analysis. This is possible simply by recognizing that a time-scaled diagram is an elaborate form of bar chart. Most computer programs will create a variety of bar charts based on the logic of the CPM network. Figure 5.15 is an early start bar chart for the highway bridge presenting the programmed schedule for the entire project. For each activity, the shaded ovals extend from its ES to EF times (black ovals represent critical activities and shaded ovals noncritical activities). The white ovals that extend to the right of the noncritical activities represent the total float. Notice that no work is scheduled for weekends or holidays. Milestones can be indicated on bar charts and frequently play an important role in the monitoring of job progress. The milestone discussed previously in Section 5.25 associated with the completion of activity 310 is shown as a triangle in Figure 5.15.

Single activities may not always be the most desirable basis for the preparation of bar charts. Simpler diagrams with fewer bars and less detail may be more suitable for high-level management. In such cases, a bar chart can be prepared using larger segments of the project as a basis. In this regard, the concept of project outlines involving different degrees of work breakdown can be valuable. This matter was discussed previously in Section 4.6. Bar charts may or may not include job restraints such as the time required for the preparation of shop drawings and for the fabrication and delivery of job materials. In general, such restraints are a function of job expediting (see Section 8.18), which is handled more or less separately from the field construction operations. Because of this, restraints that appear on the highway bridge network are not shown on the bar chart in Figure 5.15.

5.29 BAR CHART FOR REPETITIVE OPERATIONS

For repetitive operations such as the pipeline relocation, a somewhat altered form of bar chart better represents the field schedule information and is in common usage. This type of bar chart is often referred to as a line of balance chart. Figure 5.16 illustrates such a chart for the pipeline project. In this figure, the horizontal scale is

CONSTRUCTION

Project _____ Highway Bridge _____

Activity	Activity number	June 14	21	28	July 5	12
Move in	40	▨▨▨□□	□□□□□	□□□□□		
Excavate abutment #1	90	▨▨	▨□□□□	□□□□□	□□□	
Prefabricate abutment forms	80	▨▨	▨□□□□	□□□□□	□□□□	□□□□
Excavate abutment #2	120		▨▨□□	□□□□□	□□□□	□□
Mobilize pile driving rig	100	▨▨	□□□□□	□□□□□	□□□	
Drive piles abutment #1	110				▨▨▨□	□□
Forms & rebar footing #1	130				▨	▨□□□
Drive piles abutment #2	140				▨	▨▨□□
Demobilize pile driving rig	160					▨□
Pour footing #1	150					▨□□
Strip footing #1	170					▨□
Forms & rebar footing #2	190					▨
Pour footing #2	210					
Forms & rebar abutment #1	**180**					
Strip footing #2	230					
Pour abutment #1	**200**					
Strip & cure abutment #1	**220**					
Backfill abutment #1	280					
Rub concrete abutment #1	270					
Forms & rebar abutment #2	**240**					
Pour abutment #2	**250**					
Strip & cure abutment #2	**290**					
Rub concrete abutment #2	300					
Backfill abutment #2	310					
Abutments finished	315					
Set girders	**320**					
Deck forms & rebar	**330**					
Pour & cure deck	**340**					
Strip deck	**350**					
Saw joints	380					
Paint	**370**					
Guardrails	360					
Cleanup	**390**					
Final inspection	400					
Critical activities in bold						

Figure 5.15 Highway bridge bar chart schedule.

PROGRESS CHART

Job No. _____ 9108-05

		August					September	
19	26	2	9	16	23	30	6	13
□	□							
□								
□	□							
□								
□	□							
▨	□□□□□	□□□						
	▨□□□□	□□□□						
	■■■■							
	▨□□□	□□□□□						
	■■							
	■■■							
		▨▨■□□	□□□□					
		▨▨■□□	□□□□□	□□□□□	□□□□□	□		
		■■■■						
		■	■					
			■■■					
			▨	▨▨□□□	□□□□□	□		
			▨	▨▨□□□				
				⚠				
				■■				
				■■■■				
					■■■			
					■■	■		
					▨□	□□□□□	□	
						■■■■	■	
						▨▨▨□	□	
							■■■	
								■

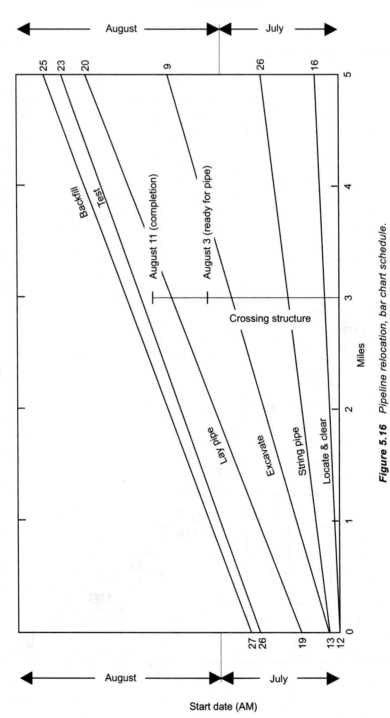

Figure 5.16 *Pipeline relocation, bar chart schedule.*

one of distance along the right-of-way while the vertical scale represents calendar time. Figure 5.16 has been prepared on the basis that the pipeline construction will be initiated on Monday, July 12. Each of the sloping lines represents the work category indicated and is plotted on an early start basis. Figure 5.10 provides the necessary ES and EF values. A conversion calendar such as that in Figure 5.5, with the numbering of working days starting on July 12, serves for the translation of these early times to calendar dates. The vertical line at mile 3 in Figure 5.16 represents the construction of the pipeline crossing structure whose beginning will also be on Monday, July 12. The crossing structure is scheduled to be ready for pipe by the afternoon of August 3 and to be completed on August 11.

5.30 COMPUTER APPLICATION TO SCHEDULING

Chapter 5 has presented a comprehensive discussion of project scheduling, a process of establishing a calendar date schedule for the field construction process. For reasons already explained, the scheduling procedure discussed herein has emphasized manual methods with the objective of developing a thorough understanding of the procedures involved and the significance of the project time data generated.

Realistic job logic and accurate activity duration estimates hopefully result from the project management team's knowledge, experience, intuition, and discerning judgment. It is important to note that the network logic and activity durations so developed provide a graphic and mathematical job model. This model is very powerful in that it allows a manager to look into the future and make project decisions based on information from the model. Like all models, the quality of the information obtained is directly proportional to the accuracy of the model itself. No amount of computer application can change this basic fact.

With regard to the mechanical process of project scheduling, however, computers enjoy a distinct advantage in their ability to make time computations accurately, with great rapidity, and to present this information in a variety of useful forms. As a result, computers are universally used for project scheduling purposes. Computers can provide the project team with an almost unlimited array of project data and graphic representations of network scheduling information. In addition to forward and backward pass calculations, the computer can convert expired working days to calendar dates. Different activities can be assigned to different calendars. For example, maintenance and support activities can be assigned to a weekend calendar while construction activities are assigned to a weekday schedule. Activity lags (Section 5.20) and hammock activities (Section 5.24) are easily incorporated into the project schedule, and a variety of activity sorts are available.

Calculations performed by computer can be displayed graphically in the form of project networks, or portions thereof, or in the form of tabular reports. These can be made to cover varying portions of the overall project and differing periods of time. Computers have the capacity to sort information in terms of specific activities, spans of time, physical location on the site, areas of responsibility, or other desired criteria. This capability eases the burden of getting the right information to the right person at the right time. Computer programs typically have the capability of reducing the network time schedule to bar chart form. These work well for reporting job progress to owners, design professionals, and others concerned with the construction schedule.

Computer programs used for project scheduling can differ substantially from one another in many important respects, and there are a wide variety of scheduling programs available. For this reason, care must be exercised to select a program best suited to the specific management needs of a given project. Most of these programs will allow a scheduled time to be assigned for the completion of construction and will make the backward pass using this designated finish date rather than the value obtained during the forward pass. Scheduled dates can also be assigned to network milestones.

6

ARROW NOTATION

6.1 PLANNING USING ARROW NOTATION

Previously, it has been stated that two symbolic conventions are used in the drawing of project networks. Precedence diagramming and the calculations associated therewith have been discussed in Chapters 4 and 5. The arrow notation system is also used by segments of the construction industry.

Planning and scheduling a construction project are basically the same regardless of the notation system used. However, the details of diagram construction and network calculations do differ between the two. The understanding and practical usage of either procedure require that the practitioner be familiar with the workings of the particular system involved. This chapter purports to acquaint the reader with how arrow notation is applied to the planning and scheduling of construction operations.

6.2 THE ARROW DIAGRAM

When arrow notation is used, each activity is depicted as an arrow, with the tail of the arrow being the starting point of the activity and the head of the arrow representing its completion. Activity arrows can be drawn to a time scale (see Section 6.18), but in the initial development of a network, at least, the lengths of the arrows have no significance. Arrows are normally drawn as straight lines or segments thereof, and their directions simply indicate the flow of work and time. The sequential relationship of one activity to another is shown by their relative positions.

Figure 6.1 shows how arrow notation compares with the now familiar precedence notation. In this figure, dashed arrows show certain kinds of dependencies with arrow notation. These arrows are called "dummy activities," and their nature is discussed in Section 6.3. Dummies are not involved with precedence notation. The accurate

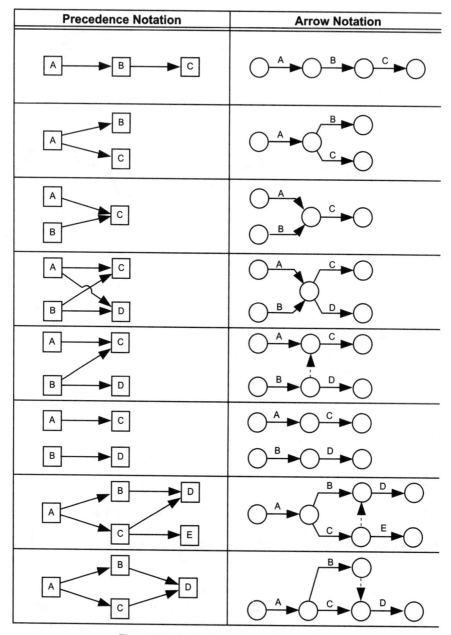

Figure 6.1 Precedence and arrow notation.

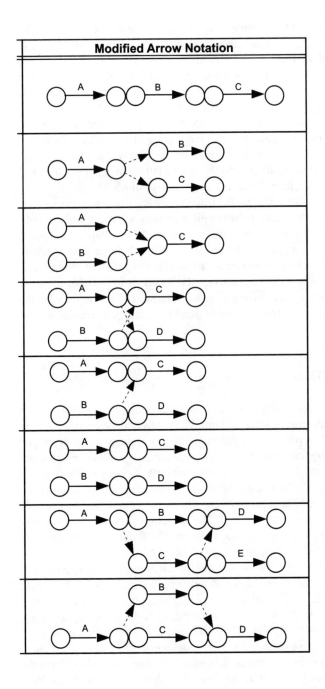

usage and interpretation of dummies is the most difficult and bothersome aspect of arrow diagramming.

To illustrate the mechanics of arrow diagramming, consider once again the construction of the pile-supported concrete footing described in Section 4.8 and precedence diagrammed in Figure 4.1. The arrow diagram depicting the same prescribed sequence of activities and containing the identical logic as Figure 4.1 is shown in Figure 6.2.

An examination of Figure 6.2 will disclose some important characteristics of arrow diagrams. Each arrow or activity is identified by the numbers located respectively at its tail and its head. For instance, the activity "Place forms" is designated as activity 50–60. In arrow diagram usage, the letter i is used as a general symbol to indicate the tail or start of an activity and the letter j to designate the head or finish. The activity "Place forms" has an i of 50 and a j of 60.

In common with precedence notation, the usual application of arrow convention requires that a given activity cannot start until all those activities immediately preceding it have been completed. To illustrate, Figure 6.2 indicates that "place rebar" cannot start until "Fine grade," "Place forms," and "Order & deliver rebar" have all been finished. In conventional arrow networks, the finish of one activity cannot overlap the start of a following activity. However, it is possible in arrow networks to depict the overlap of activities or to have the start of one activity lag the finish of an antecedent activity. This is accomplished by using "lag arrows," a topic discussed in Section 6.16.

6.3 DUMMY ACTIVITIES

In Figure 6.2, the dashed arrow 40–50 is not a time-consuming activity, but is an example of a logical connection called a "dummy" or "dependency arrow." In the case of dummy 40–50, its presence is necessary if job logic is to be portrayed correctly. The job logic stipulates that the start of "Fine grade" must await completion of only "Drive piles." "Place forms" cannot begin until *both* "Drive piles" and "Build forms" have been completed. The dummy arrow 40–50 is not an activity as such, but does show that the start of "Place forms" depends on the completion of not only "Build forms," but also of "Drive piles." The direction of the dummy arrow designates the flow of activity dependency. Dummy 40–50 does not establish any dependence of "Fine grade" on the completion of "Build forms."

The dummy 20–30 in Figure 6.2 illustrates a different usage of dummy arrows. If dummy 20–30 were not present, both "Order & deliver piles" and "Excavate" would have the same i–j designation. This circumstance would not necessarily cause great difficulty if only manual computations were involved, since the human operator can differentiate between the two activities. However, some computer programs identify an activity solely by its i–j numbers and require that each activity have a unique i–j designation. Under these circumstances, when two activities potentially have the same i–j numbers, a dummy activity is introduced as shown by dummy 20–30 in Figure 6.2. The activities "Order and deliver piles" and "Excavate" now have different i–j identities.

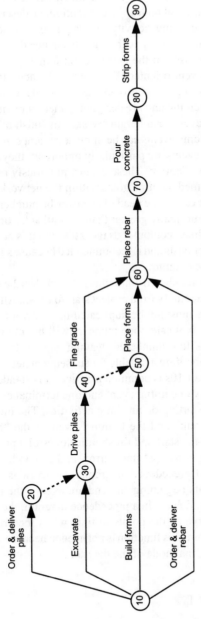

Figure 6.2 Concrete footing, arrow diagram.

6.4 NETWORK EVENTS

In Figure 6.2, the circles at the junctions of the arrows are called "nodes" or "events." All arrows must start and finish at nodes. An event might be described as that instant of time when the latest finishing activity coming into that node has just been completed. Events, as such, do not appear on a precedence diagram.

As normally drawn and used in the construction industry, arrow diagrams are activity oriented and not event oriented. This means the activities are named and emphasized in the generation and use of management data. The events are not accorded this degree of attention and are designated merely by numbers. However, the occurrence of certain events can herald the start or finish of important project segments. Consequently, some events can be of management significance and merit specific attention. If some events are particularly important, they can be named and so designated on the arrow diagram. These events, previously referred to as mile-stones, are important intermediate time goals within the network.

It is customary that the event at the head of an arrow be numbered higher than the event at its tail; that is, j is normally greater than i for all activities. Originally, this was done primarily to facilitate computer usage, although it is no longer necessary. Nevertheless, the practice persists and is recommended because such a system makes it easier to read and use the diagram.

Events are not usually numbered until the network has been completed, the numbering being done sequentially from project start to finish. All numbers need not be used, but event numbers must not be duplicated in a network. If this occurs and a computer is being used, duplicate node numbers will be interpreted as a logical loop. When the diagram is first numbered, leaving gaps in the event numbers is desirable so that spare numbers are available for subsequent network modifications. Numbering of nodes by 5s or 10s is common practice. It is standard procedure that an arrow diagram have only one initial event and one terminal event. This does not at all limit the number of starting or finishing activities. The initial event is often referred to as the "start event" and the terminal event is the "finish event," this procedure being similar to the start and finish milestones of a precedence network. Arrow diagrams do not begin with a "start" arrow and end with a "finish" arrow in a manner similar to that of precedence diagrams. An arrow is a time consuming activity and the circles at its beginning and end are events, the instants of time at which the activity begins and ends. In a precedence diagram, the initial "start" and final "finish" activities serve as the points in time when the project initiates and finalizes. In an arrow diagram, this function is not needed because it is already served by the initial node and the final node of the diagram.

6.5 DIAGRAM FORMAT

Figure 6.2 discloses that arrow diagrams, like precedence diagrams, are prepared using a horizontal network format with the project start on the left and time flow being from left to right. Project completion is at the right terminus of the diagram. A portion of each activity arrow is drawn horizontally. This portion is used for the entry of activity identification and activity time. The horizontal convention, however, does not apply to dummy activities because they are of zero time duration and are not identified other than being shown as dashed lines.

The use of backward activity arrows is not good practice and is not recommended. Such arrows act against the established network time flow and are confusing to those using the diagram. In addition, backward arrows increase the chances of unintentional logic loops being included in the network. Logical loops, previously discussed in Section 4.9, can occur in arrow as well as in precedence diagrams. Crossovers occur when one activity arrow must cross over another to satisfy job logic. These are not arrow intersections, because activities can come together only at the nodes. Any convenient symbol can be used to designate such crossovers.

Every activity in the network must have a definite event to mark its beginning. This event may be either the start of the project or the completion of immediately preceding activities. Each path through the network must be continuous with no gaps, discontinuities, or dangling activities. Consequently, all activities must have at least one activity following except those that terminate the project.

6.6 ARROW DIAGRAM OF HIGHWAY BRIDGE

Figure 6.3 is the arrow diagram for the highway bridge described in Section 4.11. Figure 6.3 depicts in arrow notation precisely the same job logic as is shown in Figure 4.3 in precedence notation. For example, dummy 210–220 in Figure 6.3 is included to show that the same forms are to be used for both concrete abutments, the forming of abutment #2 proceeding after abutment #1 has been poured and the forms stripped. In a similar manner, dummy activity 130–150 is used to indicate that the same forms will be used for the two footings.

As earlier stated, the proper use of dummies is the most difficult aspect of arrow diagramming. It is easy even for an experienced diagrammer to include false dependencies in an arrow diagram. To illustrate this, reference is made to dummies 80–90 and 80–100 in Figure 6.3. The correct dependencies in this area are (1) the start of activity 100–110 depends on the finish of activities 60–100 and 70–80, and (2) the start of activity 90–140 depends on the finish of activities 70–80 and 30–90. Figure 6.3 correctly shows this. However, if the planner had made events 80 and 100 one and the same (as, say, node 100), the diagram would not describe the established job logic because it would then indicate that the start of activity 90–140 also depends on the finish of activity 60–100. This is incorrect, because the start of activity 90–140, "Drive piles, abutment #2," obviously has no dependency on activity 60–100, "Fabricate & deliver footing rebar." Inclusion of false dependencies in an arrow diagram must be guarded against. One effective way to do this is to run through the network backward, checking the logic at each node, after the full diagram has been developed.

6.7 SCHEDULING USING ARROW DIAGRAM

The general aspects of project scheduling are essentially the same for either notation system. After the network has been prepared, an estimate of the time duration for each activity is made. Working days is the usual time unit used in this regard. Network calculations are then made. These calculations can be performed manually or by computer. Using arrow notation, it is possible to compute limiting times either for each activity or for each of the events (nodes). Although "activity times" and "event times" are defined and computed as separate quantities, they are actually closely related as will be pointed out later in this chapter.

The process of calculating activity times using arrow notation varies from that using precedence networks. Activity times can be determined manually on the arrow diagram itself, or a tabular format can be used. An alternate computational procedure uses the concept of event times. An event time is defined to be the instant in time when *all* of the activities merging into that event have been completed. For a given event, the "early event time" is the earliest possible time the event can occur considering the time requirements of those activities preceding it. The "late event time" equals the latest time the event can possibly occur and still have project completion within a designated construction period. Event computations are made directly on the diagram itself using manual methods and are an exceptionally fast and convenient way to make network computations.

6.8 ACTIVITY COMPUTATIONS FOR HIGHWAY BRIDGE

The manual computation of activity times in tabular form from an arrow diagram will be discussed first. Reference is made to Figure 6.4, the arrow diagram for the highway bridge. The time established in Chapter 5 for each activity has been entered under the activity arrow. The overall project contingency has been entered as a terminal arrow with a duration of six working days. When computing the early and late activity times from an arrow diagram, the network is used for the logic, and the calculated values are entered into a table similar to that shown in Figure 5.4. In fact, the activity times already in Figure 5.4 are exactly the same as those obtained from the arrow diagram computations. Arrow diagrams conventionally begin with a single event while precedence diagrams start with a single activity. Activity 1, "Start," in the highway bridge precedence diagram is not needed in the arrow diagram. To explain the mechanics of arrow notation computations, the numerical values contained in Figure 5.4 will now be reviewed. Activities are now referred to in terms of their *i–j* designations.

The forward pass through Figure 6.4 and Figure 5.4 is made using the same basic principles as were used in going through the precedence diagram in Figure 5.1. All computed times are in terms of expired working days as was the case with the precedence network. The ES value (column 5) for each activity in Figure 5.4 is determined first. The EF time (column 6) is then obtained by adding the activity duration to its ES. Starting at the top of column 5 in Figure 5.4, activities 10–20, 10–40, 10–50, 10–60, and 10–70 are all initial activities and their earliest possible start is zero elapsed time. Activity 10-20 has a duration of three working days, and hence the earliest it can possibly be finished is at the end of the third working day (or at a time of 3). Similarly, the early finish time for activity 10–40 is 10, for activity 10–50 is 10, and so forth.

Continuing into the network, Figure 6.4 shows that neither activity 20–30, 20–70, nor 20–160 can start until activity 10–20 has been completed. The earliest that activity 10–20 can finish is at the end of the third day, so the earliest that activities 20–30, 20–70 , and 20–160 can be started is also after the expiration of three working days (or at a time of 3). The EF of 20–30 would be its ES of 3, plus its duration of 3, or a value of 6. These values are shown on Figure 5.4.

It has just been established that the earliest finish time for activity 20–30 is six elapsed working days. The dummy 30–70 in Figure 6.4 indicates that, at least insofar as activity 20–30 is concerned, the earliest that 70–80 can start is at time 6. The ES of activity 70–80, as imposed by activity 20–70, is the ES of 20–70 (3) plus its duration (2), or a value of 5. However, 10–70 must also be completed before 70–80 can start and the EF of 10–70 is 15. It follows, therefore, that the finish of 10–70, not that of 20–30 or 20–70, controls the start of 70–80 and that 70–80 has an early start of 15. Event 70 is an example of a "merge event," this being an event with more than one activity and/or dummy coming into it. The rule for a merge event is that the earliest possible start time for all activities leaving that event is equal to the latest (or largest) of the EF values of the activities merging into it. Completing the forward pass through Figure 6.4 discloses that the EF of of the terminal activity 340–350 is 70 as is indicated by Figure 5.4.

In Figure 5.4, the calculations turn around on 70 and the backward pass is now performed. The values in columns 7 and 8 are computed concurrently, with the value of LF of each activity being determined first. In each case, the LS of an activity is then obtained by subtracting its duration from its LF value. The latest possible finish time for activity 340–350 is set equal to 70 (bottom of column 8). With a duration of 6, the latest possible start of activity 340–350 is equal to 64. The latest possible time for the finish of 330–340 is 64 since 340–350 has a LS value of 64. Commensurately, the LS for 330–340, with a duration of 1, must be 63. This same process continues in Figure 5.4 from activity to activity until the project start is reached.

Some explanation is needed when the backward pass reaches a "burst event" which is a node having more than one activity and/or dummy leaving it. In Figure 6.4, node 300 would be the first burst event reached in the backward pass. To obtain the late finish of activity 290–300, as well as activities 210–300 and 250–300, the late starts of all departing activities must first be obtained. These have already been obtained in Figure 5.4 as 55 for activity 300–320 and 57 for activity 300–310. Keeping in mind that activity 290-300 must be finished before either 300–310 or 300–320 can start, it follows that it must be finished by no later than time 55. If it were to be finished any later than this, the entire project would be delayed commensurately. The rule for a burst event is that the LF value for all activities entering that event is equal to the earliest (or smallest) of the LS times of the activities leaving that event.

This discussion of arrow diagram calculations has made no mention of dummies. In Figure 5.4 dummies could be listed as such and have their early and late times computed along with those of the regular activities. However, because dummies are of zero time duration, it is not necessary to do this. The only role that dummies play in the calculations is to indicate the activity interdependencies.

6.9   FLOATS AND CRITICAL PATH

The total float and free float values of the activities shown in Figure 6.4 are calculated in Figure 5.4 just as has already been explained in Sections 5.11 and 5.13. The critical path shown in Figure 6.4 as a heavy line is located by those activities having zero values of total float and obviously coincides with the one previously shown in Figure 5.1.

Figure 6.4 discloses that, when arrow convention is being used, the critical path can and often does include dummies. As with precedence notation, there must be at least one continuous critical path through the network from project beginning to completion. There can be multiple critical paths, either in the form of branches or as completely separate routes through the network.

6.10   EARLY EVENT TIMES

Earlier in this chapter, it was stated that when arrow notation is used, network calculations can also be performed in terms of event times, which are useful and convenient. Event time calculations are performed by hand directly on the arrow diagram. As was the case with activity times, event times are also expressed in terms of expired working days.

As defined in Section 6.4, an event occurs when all incoming activities have been finished and the early event time is the earliest that an event can occur. Values of early event times, the symbol for which is T_E, are computed by completing a forward pass through the network. Calculations still proceed in step-by-step fashion, although in this case one proceeds from event to event. As was the case with early activity times, the assumption in computing values of T_E is that each activity will start as early as possible.

Reference to Figure 6.4 will show two numbers separated by a slash mark located near each node. The number to the left is the early time for that particular event. The number on the right of the slash mark is the event late time, a topic to be discussed subsequently. Several different graphical conventions are used for the noting of event times. Squares, circles, and oval boxes are in usage. However, there is as yet no accepted standard, and the notation system used in Figure 6.4 is easy to apply and adequate for the purpose.

The computation of T_E values is simple and direct and begins at the start of the highway bridge in Figure 6.4. To start the forward pass, the early event time of node 10 is entered as zero. Because all departing activities constant at the moment this event occurs, activities 10–20, 10–40, 10–50, 10–60, and 10–70 all have an early start time of zero. Looking at activity 10–20, which has a duration of three days, it can finish at an elapsed time of three days, and its early finish time is 3. Because event 20 has only one activity (10–20) coming into it, its early event time is 3. In like manner, node 30 has only one entering activity (20–30). The T_E value of node 30 is obtained by adding the duration (3) of activity 20–30 to the T_E (3) of event 20. Thus the EF value of activity 20-30 and the T_E of event 30 are both equal to 6. The T_E value for event 70 is found as follows. Node 70 cannot occur until activity 20-30 (as indicated by dummy 30-70), activity 20-70, and activity 10-70 have all been completed. Activity 20-30 (or dummy 30-70) has an early finish time of 6. Activity 20-70 has an early finish time equal to its early start plus its duration, or $3 + 2 = 5$, and activity 10-70 will finish at time 15. Consequently, the early event time of node 70 is 15 elapsed working days. To restate a rule cited previously, the T_E of a merge event is the latest (largest) of the incoming activity EF values. The event time calculations proceed to the right in Figure 6.4 until event 350 is reached, which has a T_E value of 70.

The T_E value for a given event is the length of the longest path from the project start to that event. The early event time of a given node is also equal to the early start time (column 5 of Figure 5.4) of the immediately following activities.

6.11   LATE EVENT TIMES

The basic premise in computing late event times, symbolized by T_L, is that every event occurs as late as possible without disturbing the project completion time of 70 working days. In Figure 6.4, the first step is to indicate a T_L value of 70 for the terminal event 350. A value of T_L is entered to the right of the slash mark at each node as the backward pass progresses.

The T_L value of event 340 is determined by subtracting the duration of activity 340-350 (6 days) from the T_L value of event 350 (70), giving a value of 64 days. In a manner just opposite that of of the forward pass and repeating the process just described, T_L values of 63 and 60 are obtained for events 330 and 320, respectively. Node 310 has a late event time of 60, the same as for node 320, because only a zero time dummy separates them. At node 300, activity 300–310 will have a late start time of $60 - 3 = 57$. Activity 300–320 will have a late start time of 55. Recalling that T_L is the latest time an event may be reached without delaying project completion, it is obvious that the lesser number controls and event 300 has a T_L value of 55. If the event were to occur as late as 57, the total duration of the project would be increased to 72. The rule, again restated, is that when working backward to a burst event, the smallest of the late activity starts prevails.

As can be seen, T_L is the earliest of the late start times of the activities leaving the event and also represents the latest finish of all entering activities (column 8 of Figure 5.4). The calculations proceed backward through the network, event by event, until T_L values have been obtained for each event. The final result of this process is shown in Figure 6.4.

6.12   CRITICAL EVENTS

A study of Figure 6.4 will show that at some nodes, the values of T_E and T_L are the same. These are called "critical events" because zero time elapses between their earliest possible and latest possible occurrences. These critical events form a continuous path through the network from start to finish. Figure 6.4 discloses that this is the same critical path already located by means of critical activities with zero total float.

There are cases, however, where there is more than one choice of activity path between critical events. Alternate paths between events 300 and 320 in Figure 6.4 is such an instance. The path consisting of activity 300–320 and the other path of activities 300–310 and 310–320 both appear to be possibilities. One way to determine which path is critical is to compute the total floats of activities 300–310 and 300–320. That activity with zero total float will be the critical activity. How to compute total floats from event times is discussed in Section 6.13. However, a more immediate way to ascertain the critical activity or activities is simply to determine the longest path between the two critical events. In Figure 6.4, it is obvious at a glance that activity 300–320 is the proper choice.

6.13   CALCULATION OF FLOATS FROM EVENT TIMES

For a variety of reasons, it is useful to be able to compute values of total and free float from event times. Event times are sometimes the only network times obtained and it is important to be able to get the activity float times from them.

To find the total float of a given activity using event times, the T_E value for the event at its tail (T_{Ei}) and the T_L value for the event at its head (T_{Lj}) are determined from the diagram. Total float is obtained by subtracting the sum of the activity duration (D_{ij}) and the T_{Ei} value from the T_{Lj} value. This can be expressed in equation form as

$$TF = \text{total float} = T_{Lj} - (D_{ij} + T_{Ei})$$

The free float of an activity can be found by subtracting the sum of T_{Ei} and D_{ij} from the T_E value (T_{Ej}) for the event at the head of the activity.

$$FF = \text{free float} = T_{Ej} - (D_{ij} + T_{Ei})$$

Another category of spare time can be obtained for events by subtracting T_E from T_L. This is called "event slack" and, as has been seen, critical events have zero slack. Although event slack is related to activity floats, it is not an especially useful concept for construction applications. In the development of the original CPM procedures, the terms "float" and "slack" were maintained as separate and distinct concepts. However, present-day application often uses both terms interchangeably to refer to activity leeway. Only the word "float" is used in this text.

6.14   REPETITIVE OPERATIONS

The use of arrow notation for drawing job diagrams that involve repetitive operations has the disadvantage of requiring the usage of many dummies. This multiplicity of dummies in such a network makes its proper preparation tricky and complicates its practical application. Nevertheless, such networks are occasionally used so their preparation and usage merit some discussion.

Figure 6.5 is the arrow diagram for the pipeline relocation. It presents precisely the same logic in arrow convention as Figure 5.10 does in precedence notation. Event times have been computed and are shown in Figure 6.5. The heavy line is the critical path located by the critical events. The total time required for the pipeline work, without allowance for time contingency, is again found to be 33 working days.

6.15 MODIFIED ARROW DIAGRAM

It was mentioned previously in the preceding section that arrow diagrams for construction projects that involve a multiplicity of repetitive operations require the use of many dummies. Such project networks are difficult to prepare and to use, a condition that can be improved by reducing the number of dummies required. This can be achieved by using a modified arrow notation as shown in Figure 6.1 which eliminates the dummies between successive sections of the same category of construction work. Figure 6.6 shows such a modified arrow diagram for the pipeline relocation. This figure has been simplified by eliminating the dummies between each successive 1-mile section of Excavate and similarly between each successive mile of String pipe, Lay pipe, and Test line. Showing a circle at the start of one section immediately following the circle at the end of the preceding section is taken to be the equivalent of a zero time dummy between that finish and start. Two successive circles at the mile sections are also used for the initial Locate & clear activities and the final Backfill activities.

The essential effect of the process described above becomes especially apparent when the pipeline diagram is drawn using the procedure from the beginning. With the double nodes, one for the end of the preceding activity and one for the start of the immediately following activity of the same work type, the arrow diagram takes form quickly and easily, much like drawing a precedence diagram. The resulting network shown in Figure 6.6 is much easier to read and interpret than is the usual arrow diagram shown in Figure 6.5. Event times shown in Figure 6.6 are computed in precisely the same manner as has already been described. In this regard, it is to be noted that the pairs of adjacent events are treated no differently than two separate occurrences, one following the other at times dictated by adjacent parts of the project. Figure 6.7 shows the lag relationships, previously presented in Figure 5.9 for precedence notation, drawn using the modified arrow procedure.

6.16 SUMMARY ARROW DIAGRAM

Just as lag relationships were used to prepare a summary precedence diagram of the pipeline relocation (see Figure 5.11), a similar concept is useful in summarizing the arrow diagram shown in Figure 6.5. Such summary diagrams are of particular importance when repetitive operations are involved. Figure 6.8 is a condensation of

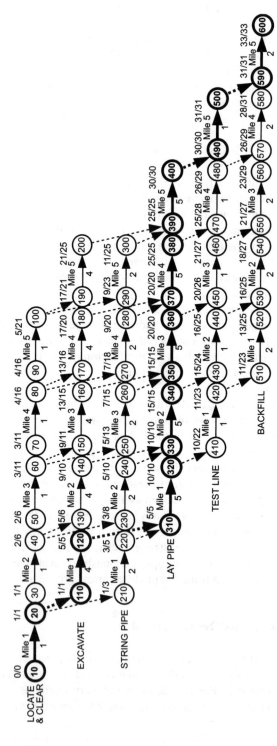

Figure 6.6 Pipeline relocation, modified arrow diagram.

Figure 6.5, showing all 5 miles of each individual operation as a single arrow. For example, activity 5–15 in this figure represents the location and clearing of the entire 5 miles of pipeline and has a total duration of five days. In a similar manner, activity 25–35 represents all the excavation and has a total time duration of 20 days. The other four job operations are shown in a similar manner. As before, excavation and string pipe are independent activities that can proceed simultaneously with one another following the location and clearing.

In Figure 6.8, the activities are represented by horizontal arrows only. All arrows directed downward are called "lag arrows." All such arrows to the left of Figure 6.8 show start-to-start relationships. In accordance with the logic of Figure 6.5, activities 25–35 and 45–55 of Figure 6.8 can begin one day after activity 5–15 starts. This circumstance is shown in Figure 6.8 by means of lag arrows (5–25 and 5–45), each with a time of one day. In a similar way, activity 65–75 can start four days after excavation starts and two days after string pipe begins. This is shown by lag arrows 25–65 (four days) and 45–65 (two days).

The downward arrows on the right of Figure 6.8 indicate finish-to-finish lags. Activity 25–35 can finish four days after activity 5–15 finishes. This fact is depicted by arrow 15–35 with a duration of four days. All the other lag arrows shown function in a similar manner. Event times have been computed in Figure 6.8 for the summary diagram and, as can be seen, the now familiar duration of 33 working days is again obtained and the same critical path is located. Figures 5.11 and 6.8 present the same summary logic for the pipeline relocation.

6.17   INTERFACE EVENTS

The pipeline crossing structure has been used previously in discussions concerning interfacing networks. Figure 6.9 is the arrow diagram of the crossing structure that interfaces with the pipeline relocation network shown in Figure 6.5. The interface between the two networks is necessitated by the job requirement that the crossing structure, which is located at the end of the third mile of the pipeline, be ready to accept pipe by the time the pipelaying has progressed that far. This construction requirement can be imposed by making event 85 in Figure 6.9 identical with event 300 in Figure 6.5. All this accomplishes, however, is to give activity 220–300 in Figure 6.5 and activities 65–85 and 75–85 in Figure 6.9 the same terminal event. Ensuring that event 85 in Figure 6.9 is reached at least by the time event 300 is achieved in Figure 6.5 is now a matter of scheduling. The event 85–300 is an example of an interface event, a common symbol for which is shown in Figures 6.5 and 6.9. The double-headed dashed arrow is not a dummy activity, and there is no flow of time or work. It merely indicates that the event concerned is an integral part of both arrow diagrams.

The essential condition of the network interface is that event 85, Figure 6.9, must occur no later than event 300, Figure 6.5. Referring to event 300, it can be seen that it has a T_E value of 20. This means, then, that event 85 must be given a T_L of 20. The late event time of 20 has been entered at event 85 of Figure 6.9, and T_L values have been computed back to the beginning of the network. This backward pass discloses

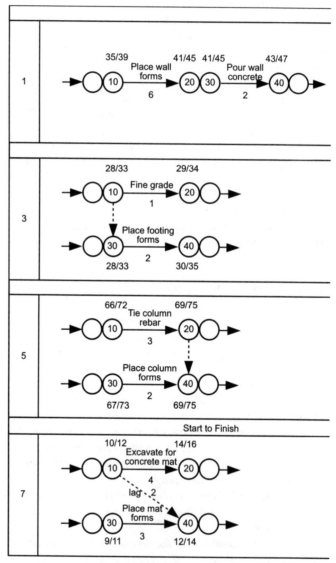

Figure 6.7 Lag relationships, modified arrow notation.

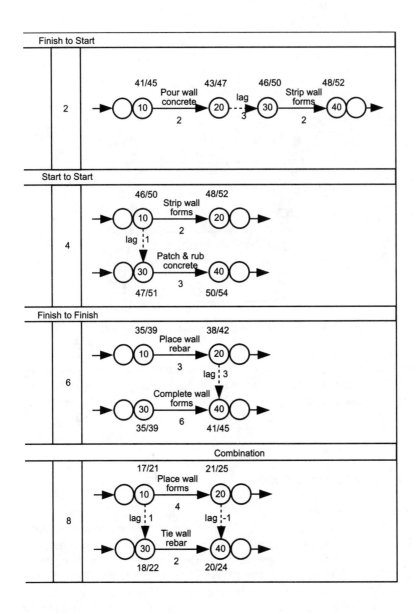

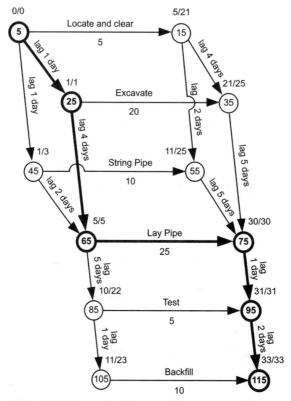

Figure 6.8 Pipeline relocation, summary arrow diagram.

that the T_L of the beginning event (5) is 3, signifying that construction of the crossing structure must be started no later than the beginning of the fourth day (time = 3). Making a forward pass using a T_E of event 5 equal to zero yields the early event times shown in Figure 6.9. The significance of the event times obtained has already been discussed in Section 5.23.

6.18 TIME-SCALED NETWORK

The drawing of a time-scaled network was discussed in Section 5.26, a time plot of the highway bridge being obtained from its precedence diagram. When an arrow diagram is the source network, a time-scaled plot is prepared in much the same manner. Figure 5.14, which is the time plot of the highway bridge, would be obtained from Figure 6.4, the arrow diagram of the highway bridge, as follows. The early start and early finish for each activity is plotted to an established horizontal time scale. The length of the connecting arrow is the duration of that activity. Events are also plotted at their T_E times, being indicated by circles. The free floats of individual activities are obtained automatically from the resulting plot. The resulting figure obtained from

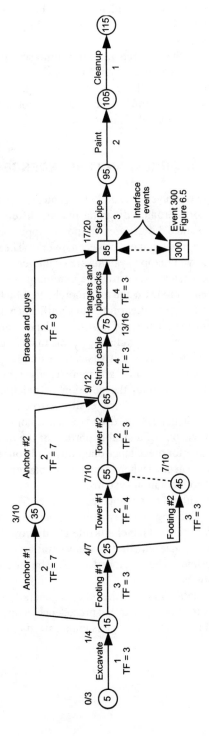

Figure 6.9   *Pipeline crossing structure, arrow diagram event time computations.*

the arrow diagram would be essentially identical with Figure 5.14. One difference might be that the events would be numbered as they are in Figure 6.3 and the activities indicated by their *i–j* designations.

An argument can be made that the preparation of a time-scaled figure is intuitively more obvious from an arrow diagram than it is from a precedence network. To some, the time plot is a natural extension of the usual arrow diagram, but is not a logical extrapolation from precedence notation. In any event, the obtaining of a time-scaled diagram from either source is not a difficult process.

6.19   COMPARISON OF ARROW AND PRECEDENCE NOTATION

Even though the job logic portrayed is exactly the same, the arrow and precedence methods of drawing networks produce diagrams that are different in appearance. A comparison of Figures 4.3 and 6.3 illustrates this point. Appearance, however, is not of concern to those who originate and use project networks. The relative advantages and disadvantages of the two notation systems are of interest to job management.

The advantages of precedence over arrow diagramming are several. Precedence notation is considerably easier to learn and to comprehend. Much of the explanation for this is the absence of dummies. Learning the significance and proper usage of dummies requires time and experience and false dependencies are a very real hazard with arrow diagrams. Precedence diagramming seems to be intuitively simpler and more obvious to the neophyte because it is basically a simpler procedure. Another advantage is that precedence diagrams are more easily revised than an arrow diagram. Using precedence notation, the addition or deletion of activities and changes of logic can be accomplished fairly easily. Arrow diagrams, on the other hand, are relatively inflexible in this regard, and even a minor network modification is apt to involve major revisions of a sizable portion of the diagram. The activity numbering system is simpler for precedence diagrams. Only one number is involved while two numbers are often needed when using arrow notation. Adequate computer software is available for both notation systems.

Arrow notation is not without its advantages, however. The calculation of event times is an exceptionally advantageous way to make manual network computations. Those familiar with the workings of bar charts often find arrow diagrams to be more meaningful and easier to associate with the time flow of job activities. Arrow notation was the first of the two systems to be developed and used by the construction industry for job planning applications.

The adoption of one notation mode or the other is generally a matter of management choice. On balance, it would appear that precedence diagramming is the better system of the two. For this reason, its usage is emphasized in this text.

7

PROJECT TIME
ACCELERATION

7.1 TIME SCHEDULE ADJUSTMENTS

Project work schedules must often be adjusted to accommodate adverse job circumstances. Revisions of the project schedule are commonly required so that contract
time requirements can be met. Established time goals can make it imperative that key
stages of the work be achieved earlier than originally planned. The start or finish
dates of major job elements often have to be improved to satisfy established time
constraints or commitments. Milestones, network interfaces, and final completion
are common examples of key events that sometimes have to be rescheduled for earlier
dates. Such schedule advances are accomplished, in practice, by performing certain
portions of the work in shorter times than had been originally allocated to them.

This chapter discusses some of the reasons why management action to reduce
project time is occasionally needed and how the associated studies are conducted.
The highway bridge will be used for purposes of discussion and illustration.

7.2 NEED FOR TIME REDUCTION

There are many practical examples where shortening the time of selected job
elements can be desirable as a means of meeting important project target dates. For
instance, the owner may impose by terms of the construction contract a job
completion date that the present project plan will not meet. Failure to meet this
contractual time requirement will put the contractor in breach of contract and make
it liable for any damages suffered by the owner because of late project completion.
On a job in progress, the owner may desire an earlier completion date than originally
called for by the contract and request that the contractor quote a price for expediting

the work. It is entirely possible that the programmed project duration time may not suit the contractor's own needs. The contractor may wish to achieve job completion by a certain date to avoid adverse weather, to beat the annual spring runoff, to free workers and equipment for other work, or to meet other such circumstances. Financial arrangements may be such that it is necessary to finish certain work within a prescribed fiscal period. The prime contractor may wish to consummate the project ahead of time to receive an early-completion bonus from the owner. A common motivation for time acceleration is where the work is well underway and delays have resulted in a substantial loss of time that must be recovered by the end of the project.

Although not involving the entire project, a similar situation can arise in attempting to meet an established milestone. It is not unusual that the computed early times of milestone events occur later than desired. An entirely analogous situation can arise with respect to network interface events.

This clearly discloses one of the great advantages of being able to establish advance construction schedules of reasonable accuracy. Such information makes it possible for the project manager to detect specific time problems well in advance and to initiate appropriate remedial action. Certainly, this is preferable to having no forewarning of such problems until it is too late to do much, if anything, about them.

On first impression, it may appear incongruous to consider shortening the duration of a project when a contingency allowance has been added to the computed normal time. In the case of the highway bridge, a contingency of six days was added to the calculated time of 64 days to establish a probable completion time of 70 working days. This project time is the best advance estimate available of the work duration that will actually be required. As a consequence of this fact, 70 working days is the most likely duration of the highway bridge, and any need to shorten the project should depend on how this time compares with the prescribed contract period or other established time limitation.

7.3   GENERAL TIME REDUCTION PROCEDURE

Perhaps it is appropriate now to mention that a variety of terms are applied to the process of shortening project time durations. "Least-cost expediting," "project compression," and "time-cost tradeoff" are all used in reference to the procedures discussed in this chapter. The exact nomenclature is not especially important so long as the method and its application are thoroughly understood. The use of the term "expediting" in connection with shortening project time requirements is unfortunate because the same term is also used with regard to actions taken to ensure timely resource support for construction operations. However, the double meaning of expediting is commonplace in the construction industry and is so applied herein.

To shorten the time period required to reach a milestone event, interface event, or to achieve project completion, one need be concerned with reducing the time durations of only a certain group of activities. As has already been shown, the time required to reach any future network event, terminal or otherwise, is determined by the longest time path from the present stage of project advancement to that event. Consequently, if the time to reach a certain event is to be reduced, the only way this can be accomplished is to shorten the longest path leading to that event. This is a very revealing and important observation. In the absence of such management information, the usual reaction when a project is falling behind schedule is to expedite, willy-nilly, all the ongoing activities in an attempt to make up the lost time. The inability to discriminate between those activities that truly control and activities of little or no consequence timewise can make such expediting actions far more expensive than is necessary.

When the date of project completion is to be advanced, it is the network critical path that must be shortened. In the case of reducing the time to achieve a milestone or interface event, the diagram critical path itself may not be involved in any way. Insofar as that particular event is concerned, it is the longest time path leading to it that must be shortened, and this path may be entirely separate from the critical path which applies for the entire network. For purposes of discussing the reduction of project time, the longest time path leading to the event in question will be referred to as its "critical path."

At this point, it must be recognized that when a longest path is shortened, the floats of other activity paths leading to the same event are commensurately reduced. It is inevitable, therefore, that continued shortening of the original critical path will lead, sooner or later, to the formation of new critical paths and new critical activities. When multiple critical paths are involved, all such paths must be shortened simultaneously if the desired time advancement of the event is to be achieved. Shortening one critical path, but not another, accomplishes nothing except to provide the shortened path with unneeded float.

When time reduction is done manually, the effect of each shortening action must be checked to ascertain if it has produced new critical activities. The usual way to do so is to perform a network recomputation after each step in the time reduction process. Such recomputations can be done manually or by computer. If the network is not large, manual computations can be fast and convenient. For a large network, many successive recalculations can become a substantial chore. Even though time-scaled diagrams are of limited value in making actual time reduction studies, such plots are especially useful for explaining the total effect of a given time reduction. A network change can be visualized in terms of movements of rigid-frame portions of the diagram and the effect of these movements on its elastic (float) connections. Whenever any shortening action eliminates a dashed float line in a time-scaled network, a new critical path is automatically formed. Portions of Figure 5.14 are used in this chapter for the purpose of giving the reader a better appreciation of what network time reduction actually involves.

7.4 SHORTENING THE LONGEST TIME PATH

It has now been established that if the date of a specified project event is to be advanced, the length of the longest time path leading to the event must be shortened. Basically, there are only two ways to accomplish this. One is to modify the job logic in some way such that the longest route is diminished in length. This involves a localized reworking of the original job plan, with time being gained by rearranging the order in which job activities will be accomplished. This time reduction procedure does not reduce the durations of activities themselves; it gains time by the more favorable sequencing of selected job operations.

The other possible way to reduce the length of a critical path is to reduce the duration of one or more of its constituent activities. Each critical activity must first be examined to see if any shortening is possible. The compression of an activity can be achieved in a variety of ways, depending on its nature. Additional crews, overtime, or multiple shifts might be used. It may be possible to subcontract it. More equipment might be temporarily brought in and assigned to that activity. Earlier material deliveries might possibly be achieved by authorizing the fabricator to work overtime, by using air freight or special handling, or by sending one of the contractor's own trucks to pick up and deliver the material. There usually are, of course, some activities whose durations cannot be reduced in any feasible way.

7.5 PROJECT DIRECT COSTS

Before the discussion of project time reduction can proceed further, it is necessary to discuss the nature of project direct costs and indirect costs. Costs are necessarily involved with time reduction because construction expense is a function of time. Although management discretion may occasionally dictate otherwise, an effort is usually made to achieve gains in time at the least possible increase in project cost. If project management is to make schedule adjustments at the least additional cost, then it is necessary to understand how the costs of construction operations vary with time.

The direct cost of an activity is made up of the expense of labor, equipment, materials, and subcontracts. Each activity has its normal cost and normal duration. The normal cost is the least direct cost required to accomplish that activity and is the cost customarily ascribed to the work when the job is being estimated. The normal duration is the activity duration determined during the scheduling phase. Although there is nothing precise about a normal time, it still constitutes a reasonably distinct datum or reference point for accomplishing an activity at least direct cost. It is obvious that the direct cost of the total project is equal to the sum of the direct costs of the individual activities and the normal project duration is derived from the activities' normal durations. It follows, therefore, that the least total direct cost of the entire job is that associated with the normal project duration.

If the estimated activity times can be accepted as going hand in hand with minimum direct cost, then any variation in an activity time from that estimated, either more or less, must result in a commensurate increase in its direct cost. The degree to which this rule actually applies in practice is considerably more certain with respect to increased costs caused by reducing the activity duration than by extending it. However, contractors are seldom concerned with "stretching-out" or deliberately extending the duration of a job activity.

The practical fact that shortening an activity time will normally increase its direct cost is not difficult to demonstrate. The use of multiple shifts or overtime work obviously entails extra labor expense. Crowding in more work crews or pieces of equipment makes job supervision difficult, reduces operational efficiency, and increases costs of production. Early material delivery requires payment of premiums to the vendor or increased transportation and handling costs. All this leads to the conclusion that, if a reduction is to be made in a project time requirement, the direct costs of the activities actually shortened will usually be increased.

7.6 VARIATION OF ACTIVITY DIRECT COST WITH TIME

The direct costs of activities can vary with time in many ways, although it is considered here that these costs always vary in inverse proportion with time. A continuous linear, or straight line, variation of direct cost with time is a common example. This is the result of an expediting action, such as overtime or multiple shifts, where the extra cost for each day gained is just about constant. Figure 7.1a is such a case where the normal activity time of 15 days can be reduced by as much as three days. The increase in direct cost of this activity is a constant $100 per day, this being termed the "cost slope" of the activity. The contractor may elect to expedite the activity by one, two, or three days, for which the extra cost will be $100, $200, or $300, respectively.

The "normal points" in Figure 7.1 represent the activity normal times and normal direct costs discussed before. The expediting of an activity is often called "crashing." As indicated in Figure 7.1, the minimum time to which an activity can be realistically reduced is called its "crash time," and the corresponding direct cost is called its "crash cost." The plotted intersection of these two values is called its "crash point."

Figure 7.1b shows a continuous, piecewise linear, time-cost variation. The figure indicates that a time reduction of one, two, or three days is possible, but the cost slope increases for each additional day gained. An example of a piecewise linear variation would be when crew overtime is involved and the use of a progressively larger crew is possible as the work advances. Figure 7.1c is a noncontinuous or gap variation where the expediting action reduces the activity duration from 15 to 12 working days, with no time possibilities in between. The activity is reduced by three days or none, and the extra cost of the shortening is a fixed sum of $300. This type of time-cost variation is common in construction. Paying a premium for an early material delivery or shipping by air rather than motor freight are examples.

Other forms of time-cost variation are obviously possible. However, great accuracy in determining the extra costs resulting from expediting actions is seldom achievable, and the time-cost variations used along with the expediting of construction activities are generally limited to the three contained in Figure 7.1.

7.7 PROJECT INDIRECT COSTS

As described in Section 3.21, project overhead consists of indirect costs incurred in direct support of the field work, but which cannot be associated with any particular physical portion of the job. Figure 3.7 discloses that, on the highway bridge, time variable job overhead expense was estimated to be $25,675. Time constant overhead

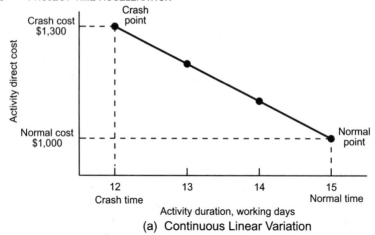

(a) Continuous Linear Variation

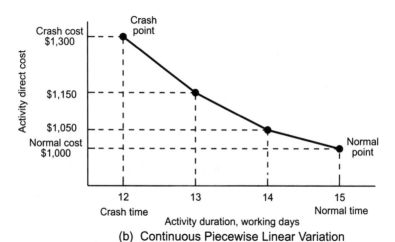

(b) Continuous Piecewise Linear Variation

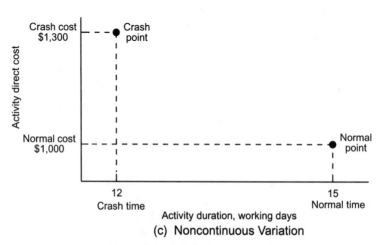

(c) Noncontinuous Variation

Figure 7.1 *Activity time-cost variation.*

expenses are not variable with project duration and need not be considered here. The probable duration of the highway bridge has been determined to be 70 working days. This means that the time variable indirect expense on this job amounts to $25,675 ÷ 70 ≈ $367 for each working day.

The preceding discussion discloses that the variation of project indirect costs is just the opposite that of activity direct costs. If an activity is shortened, its direct expense increases. If the project is reduced in length, the job indirect expense decreases.

7.8 TIME-COST TRADEOFF BY COMPUTER

A number of computer programs are available that will accept activity time-cost data such as that in Figure 7.1 and produce an optimum schedule of project cost and time. The project duration is decreased below its normal value one day at a time. For each reduced duration, the computer searches out the combination of activity shortenings that will accomplish the time reduction at least increase in project direct cost. To the increased direct cost is added the time constant overhead expense and the time variable overhead cost at the established per diem rate. This gives the total project cost for the reduced duration. The basic objective of these computer programs is to determine the project duration for which the total project cost is a minimum.

Because of major failings, however, such computerized time-cost tradeoff has found limited acceptance in the construction industry. The job models used in such analyses are grossly oversimplified and provide only "paper" solutions that have little association with the realities of a construction project. The only time-cost data that the computer can handle are time-cost slopes for the individual activities. It has no capability of optimizing time shortening when changes of network logic are involved as possible alternatives. Stated another way, the computer can only perform a time-cost tradeoff analysis by compressing individual activities in a network with a set system of logic. This is a significant limitation when it is recognized that revisions in the programmed job plan frequently account for the most significant time reductions.

The conventional computer analysis goes through a step-by-step process of expediting activities in the ascending order of their cost slopes. This unthinking and invariable insistence of least additional cost is often not in the best interests of good job management. For various reasons, some activities are more attractive choices for expediting than others on purely practical grounds. For example, the machine will select an activity for expediting on the basis of its least additional cost, even though it may be scheduled several months hence. A project manager would probably be better advised to expedite a critical activity near at hand, even at a somewhat higher incremental cost. Future uncertainty is such that the distant activity may subsequently become no longer critical or would not have provided the time reduction ultimately required. The reduction of project time must, of necessity, involve considerations other than just minimizing extra cost. These considerations, although possibly subtle or indirect, can often be of overiding importance to project management.

At this point, it suffices to say that the computer does not normally serve as an adequate stand-alone device for project shortening. Manual methods, relying on human insight and judgement, continue to play a commanding rode in the process. The project time acceleration procedures discussed herein describe and emphasize such an approach. This does not say the computer does not play an important supporting role since management and computer can work together to achieve the best solution possible. The manager can originate and pass on matters of judgment and the computer can process the decisions made by project management.

7.9 PRACTICAL ASPECTS OF TIME REDUCTION

The least-cost shortening of actual construction networks can become enormously involved. Multiple critical paths can appear that make the shortening process a very complicated procedure. The number of possible expediting combinations to be tested if an optimal solution is to be achieved can become very large. It must be recognized, therefore, that the usual manual time reduction will certainly not always provide project management with truly optimal expediting combinations. However, mathematical precision with imprecise data is certainly not the only, nor necessarily the most important, consideration involved in such a process. Time-cost tradeoff performed by computer is basically unrealistic in the sense that the management information generated does not adequately reflect the application of human judgment. The actual accomplishment of time reduction in practice must, of necessity, be concerned with a number of practical considerations other than just the matter of buying the most time for the least money.

Manual solutions for project time reductions, while perhaps not optimal, do provide invaluable guidance to the project manager in making decisions about whether expediting is practicable and, if so, how to proceed. In most cases, guidance on how to make intelligent choices of time reduction actions is as valuable as a theoretically optimal solution. Input data are uncertain, conditions change from day to day, and construction is simply not an exact or a completely predictable process. Even the critical path of a given network may change its routing occasionally as the work progresses. Project managers strive to find practical, reasonable answers, not achieve perfection. Expediting a project manually makes it possible to inject value judgments into the process and affords the project manager with an intuitive feel for the effect of expediting actions on other aspects of the projects. In addition, a project time reduction study can easily include the critical evaluation of time gained by revisions in job logic as well as by shortening individual activities.

The manual accomplishment of a project time reduction is entirely directed toward reducing the length of the applicable critical path or paths. This is a step-by-step process using time reduction measures that are considered to be feasible and best suited to the job context. These may be either changes in job plan or the shortening of individual activities or both. The usual procedure is to gain each increment of time with the least increase possible in direct cost. Where other job factors are of greater importance than incremental cost, shortening steps are taken in whatever order project management believes is in the overall best interest of the work.

7.10 REDUCTION OF HIGHWAY BRIDGE DURATION

For purposes of illustration, suppose that the contractor on the highway bridge determines that the probable duration of 70 working days or 98 calendar days is unsatisfactory. Work on this job has not yet begun, and a study is to be made concerning the feasibility and attendant cost of reducing the overall project duration by perhaps as much as 10 percent.

The essential question is, of course, how can the project critical path be reduced from its present time duration. As a starting point, common sense would surely suggest that a second look be given to the present operational plan with the objective being to possibly gain time at no increase in direct cost. Although the present job plan was carefully devised by knowledgeable and experienced people, perfection can scarcely be expected from the original planning effort. It seems likely that a restudy of the operational sequence could sharpen the planning approach and perhaps indicate opportunities for greater time efficiency.

If reexamination of the original job logic does not produce the desired time gain at no additional cost, then the contractor has no option but to pay for it. Almost any construction operation can be performed in less time if someone is willing to pay for the additional expense. There are undoubtedly a number of opportunities for shortening the duration of the highway bridge by effecting changes in job logic or by reducing the times needed to accomplish individual critical activities. When extra cost is involved, project shortening is achieved by evaluating the feasible alternatives and normally adopting the least-cost combination of those that will produce the desired time adjustment.

It is the intent of the following five sections to present specific discussions of how the overall duration of the highway bridge might be reduced at little or no additional direct cost to the contractor. Such a reexamination of the programmed plan will not always result in the gain of any time, but the possibility is there and should be investigated. A word of caution is in order, however. When such restudies are being made to pick up some badly needed time, there is always a tendency to become optimistic about things. Those who make decisions concerning project time reduction must be sensible and pragmatic in their judgments.

7.11 RESTUDY OF CRITICAL ACTIVITY DURATIONS

There is one obvious check to be made initially when reexamining the critical path of a project to be shortened: review the time estimates of the individual critical activities. Errors can be made, and it is worthwhile to verify the reasonableness of the time durations originally estimated.

There is another possibility that can be reviewed. When the time estimates were first made, it was not known which of the activities would prove to be critical. The original time estimates of some activities may have been made in contemplation of limited future availability of labor crews or construction equipment. The result of this could have been that some of the activity duration estimates for what later turned out to be critical work items were based on smaller than optimum-sized crews or equipment spreads. Now that the identities of critical activities have been established, it may be feasible to defer action on some noncritical activities, using their floats for this purpose, and reassign resources temporarily to the critical activities, with the objective of accomplishing them in more normal lengths of time.

7.12   RESTUDY OF PROJECT PLAN

A project restudy aimed at gaining time at no additional direct cost is essentially a critical second look at the established operational plan. The objective, of course, is to devise a reworking or refinement of the logic of a limited area of the network that will result in a shortening of the critical path. Here is a time when innovative thinking and a fresh approach can sometimes result in important improvements of work methods. Traditional and established field procedures are too often accepted as being the only way to get the job done and are not challenged often enough by inquiring minds who believe that a better way may be possible. A good old-fashioned brainstorming session will occasionally produce some ingenious ideas concerning new approaches.

There are times when the contractor has the authority to make changes in project materials or design or perhaps can do so with owner approval. A design-construct contract would be an instance of where the contractor, to advance a project date, might decide to make a design change that would make the reduction of a job time duration possible. Under ordinary circumstances, however, the contractor is not empowered to do this.

7.13   CRITICAL ACTIVITIES IN PARALLEL

As an example of a change in project plan made to shorten the critical path, it may be possible to do certain critical activities in parallel with each other rather than in series. To show how this might work, reference is made to the time-scaled diagram of the highway bridge in Figure 5.14. Study of the critical activities will disclose that there is the possibility of scheduling the painting to be done concurrently with stripping of the deck forms rather than afterward. Figure 7.2, which is excerpted from Figure 5.14, illustrates this change in the project schedule, showing that activity 370, "Paint," could start at the same time as activity 350, "Strip deck," thus shortening the critical path by two days. If this were done, painting would no longer be critical, but the guardrail installation would become so as indicated by Figure 7.2.

The possibility of accomplishing activities 350 and 370 in parallel with one another might generate an unsafe working condition. This would have to be resolved before this means of shortening the critical path could be approved for implementation in the field.

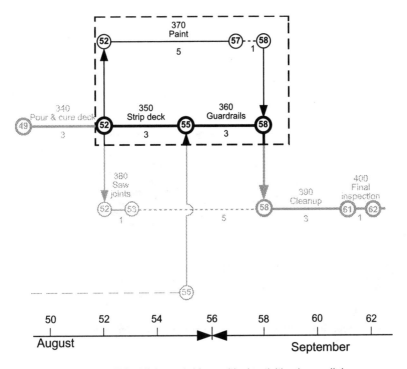

Figure 7.2   *Highway bridge, critical activities in parallel.*

7.14   SUBDIVISION OF CRITICAL ACTIVITIES

In shortening a critical path, a check can be made to determine if each critical activity must necessarily be completed before the next one can start. The judicious subdividing of one critical activity may enable it to overlap another. In other words, it may be possible that a critical activity can be subdivided with a portion of it being done in parallel with another critical activity. There appear to be two or three such opportunities in Figure 5.14, only one of which will be discussed here because of the similarity between them. Referring to activity 180, "Forms & rebar, abutment#1," one side of these forms must be erected before the steel can be tied. Consequently, the finish of activity 60, "Fabricate & deliver abutment & deck rebar," could overlap the start of activity 180 by one day, assuming that the time necessary to erect one side of these prefabricated forms is one day. Figure 7.3 shows how the original time-scaled diagram (Figure 5.14) would have to be modified to reflect this change. As can be seen, this alteration would shorten the critical path, and hence the time to any succeeding critical activity, including the terminal activity, by one day.

As said before, the possibility of other critical paths being formed must be investigated each time the critical path is shortened. For the time reduction discussed in the previous paragraph, reference to Figure 5.14 shows that all activities in the network to the right of activity 180 have moved as a unit one day to the left. In so

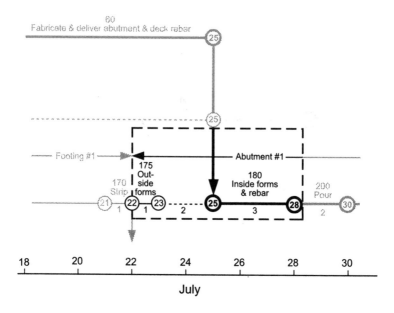

Figure 7.3 *Highway bridge, subdivision of critical activity.*

doing, the floats shown at the ends of activities 230 and 260 have been reduced by one day. In addition, the creation of the new activity 175, "Outside forms, abut. #1," reduces by one day the float following activity 80 and that following activity 170. This action does not, however, result in the formation of any new critical activities.

7.15 SUBCONTRACTING

Another possibility of shortening a project critical path might be to subcontract certain work that the general contractor originally intended to do with its own forces. The project plan may show certain critical activities to be in series with one another, not because of the physical order in which the work must be done, but because they require the same limited resource. Subcontracting all or a portion of the work involved to a specialty contractor who has adequate equipment and manpower might enable the activities to be performed concurrently rather than one after the other, thus saving considerable time.

It is not unusual for an equipment or labor restraint, together with a dependent activity, to be a part of the critical path. This usually represents a limitation on the availability of a general contractor resource. The prime contractor might consider subcontracting that part of the work to a firm whose labor crews or equipment would be available at an earlier date than its own. Whether the work could be subcontracted for the same cost that the prime contractor estimated would have to be determined. If extra cost is involved, this becomes an expediting action that will have to be considered on a comparative cost basis with the other additional expense possibilities.

7.16 TIME REDUCTION OF HIGHWAY BRIDGE BY EXPEDITING

The preceding five sections have discussed how a critical reexamination of the original job plan could possibly result in the shortening of a project critical path at no increase in direct cost. If such a network study does not produce the desired time reduction, then project management must literally "buy time" by resorting to expediting actions that will increase direct costs. Expediting actions are thus distinguished, herein, from time gained at no extra expense.

The sections following consider the expediting process as it is applied to the highway bridge. For clarity, the ensuing discussion assumes that none of the possible time reduction measures discussed previously that might have been achieved at no additional direct cost has been adopted. Therefore, the original plan and schedule of the highway bridge remain unaltered. Shortening of this project by subcontracting is not considered to be either feasible or desirable, so the ensuing time reduction measures are limited to those the prime contractor can achieve with its own forces.

The first step in expediting the highway bridge probably should be to determine if there are any changes in job plan that would shorten the critical path. As has already been stated, modifications of network logic are often more fruitful than shortening individual activities. An obvious change that could be made in the logic of the highway bridge which involves the critical path would be to prefabricate and use two sets of abutment forms. It will be recalled that the job plan for this project, as presented in Figure 5.1, calls for only one set of forms to be built, these forms being used first for abutment #1 and then reused on abutment #2. Consequently, according to the original plan, the start of the second abutment will have to await stripping of the forms from the first. A second set of abutment forms would eliminate the dependency line from activity 220, "Strip and cure, abut. #1," to activity 240, "Forms and rebar, abut. #2." This would enable the forming of abutment #2 to start just as soon as footing #2 is finished (activity 230). The effect of this change can be checked by a recomputation of the altered network or by reference to the time-scaled plot in Figure 5.14. In either case, if this change is made, the entire project will be shortened by six working days, from 70 to 64. The abutment #2 string of three activities will no longer be critical, the critical path now being routed through activity 280.

A preliminary examination of the critical activities of the highway bridge discloses that there might be several possibilities for expediting individual activities. Suppose that careful study, however, reveals that there are actually only four activities for which some expediting is considered to be practicable in the view of project management. Figure 7.4 lists the time reduction and additional direct cost for each of the ways in which the highway bridge may feasibly be expedited. Assuming that all these alternatives are equally acceptable to project management, the process is now one of shortening the project, one step at a time, each increment of time being gained at minimum additional direct cost. Project management can now decide how much time reduction it is willing to purchase.

Expediting Action	Time Reduction (Working days)	Direct Cost of Expediting
(A)		
Prefabricate two sets of abutment forms. Start placing Forms & rebar, abut. # 2 (activity 240) directly after Strip foot. # 2 (activity 230)	6 (logic change)	$3,236
Computations		
One set of forms as estimated (Figure 3.6)		
Material $1,645		
Labor $2,907		
$4,552		
Two sets of forms		
Material		
Plyform: 10% waste, 1 use, 70% salvage		
3620 (1.10)($0.38)(.30)= $454		
Lumber: 4 fbm per sf., 1 use, 70% salvage		
3620 (4)($0.35)(.30)= $1,520		
Labor		
3620 (1.10) + 46% indirect= $5,814		
$7,788		
Cost of expediting=$7,788-$4,552=$3,236		
(B)		
Steel fabricator agrees to work overtime and give abutment rebar special handling. Deliver abut. & deck rebar (activity 60) reduced from 15 to 11 days.	4	$1,400 (quoted by rebar vendor)
(C)		
Expedite Pour abutment #1 (activity 200) by bringing in an additional crane, hiring two more laborers, and working the concrete crew 2 hours overtime. Duration reduced from 2 days to 1 day.	1	$425
Computations		
Pour abutment #1 as estimated (Figure 3.6)		
Labor $3,188		
Equipment $1,715		
$4,903		
Expedited Pour abutment #1 (10 hours)		
Labor		
Regular crew $1,594		
Extra laborers and operator $828		
Overtime pay $726		
Equipment		
Cost of two cranes, etc. $1,720		
Extra crane, in and out $460		
$5,328		
Cost of expediting=$5,328-$4,903=$425		
(D)		
Expedite Pour abutment #2 (activity 250) in the same way as abutment #1.	1	$425
(E)		
Expedite Strip deck forms (activity 350) by hiring three more laborers and working 4 hours overtime. Duration reduced from 3 days to 2 days.	1	$260 (not shown but similar to previous calculations)

Figure 7.4 Highway bridge, direct costs of expediting actions.

7.17 LEAST-COST EXPEDITING OF HIGHWAY BRIDGE

Using the information summarized in Figure 7.4, the highway bridge is now to be reduced in duration with each successive increment of time compression being realized at a minimum increase in project direct cost. As each step is taken, a check must be made to determine whether that expediting action results in the formation of any new critical activities. This check is made by means of a network recomputation after each time reduction step. Figure 7.5 summarizes the results of the successive expediting actions.

Examination of Figure 7.4 discloses that the cheapest first step in the shortening process is the expediting by one day of critical activity 350 at an additional direct cost of $260. In a network with a single critical path, as is the case with the highway bridge, the amount of any step decrease in the duration of a critical activity is subject to two limitations, one internal to the activity and the other external. The first of these is how much internal shortening of the activity is physically possible. In the case of activity 350, the physical limit has been established as one day. The second limitation is based on how much the activity can be shortened before a new parallel critical path is formed. This is often referred to as the external or logical limit of an activity shortening. The logical limit of a given critical activity is equal to the total float of the shortest alternative path around that activity. Reference to Figure 5.1 shows that the path through activity 380 (TF = 7) is the shortest way around activity 350. Hence the logical limit of activity 350 is 7, which just says that activity 350 could be shortened by as much as seven days before a new critical path is formed. Obviously, the first step in expediting the highway bridge using activity 350 is limited to the lesser of its physical limit (one day) or its logical limit (seven days) or a shortening of one working day. This information is summarized in step 1 of Figure 7.5a Activities 200 and 250 are expedited in a similar manner and are shown as steps 2 and 3 in Figure 7.5a. In this figure, whether a new critical path is formed by a specific shortening step is indicated by "yes" or "no" entered in the column headed "New Critical Path."

To shorten the highway bridge to its full potential at the least additional cost, it is necessary to conduct three separate sequences of shortening actions. Figure 7.5a is the first. The successive shortening by actions E, C, and D, as described in Figure 7.5a, shortens the project by three days at a total extra direct cost of $1,110. If the project is to be shortened by only three days, this is the least expensive way to accomplish it. To shorten the project additionally, an entirely new and different sequence of shortenings must be used. In other words, the time reduction process must be started anew. Step 1 of Figure 7.5b shows that shortening activity 60 can reduce the project by three days, from 70 to 67 working days, at an additional cost of $1,400. Actually, the expenditure of $1,400 shortens activity 60 by four days, which is its physical limit. However, when activity 60 is shortened by three days, a new branch on the critical path is formed. Hence the logical limit of this shortening action is three days and shortening activity 60 only results in shortening the project by three days. If the fourth day of shortening of activity 60 is to become usable, some activity on the new critical branch will have to be shortened by one day also. Study

Step	Project Duration (Working days)	Expediting Action (from Figure 7.4)	Critical Activity Shortened	New Critical Path	Add'l. Direct Cost	Cumulative Direct Cost
(a)						
*1	69	E	350	Physical limit=1 Logical limit=7 No	$260	$260
*2	68	C	200	Physical limit=1 Logical limit=8 No	$425	$685
*3	67	D	250	Physical limit=1 Logical limit=6 No	$425	$1,110
(b)						
1	67	B	60	Physical limit=4 Logical limit=3 Yes	$1,400	$1,400
*2	66	E	350	Physical limit=1 Logical limit=7 No	$260	$1,660
*3	65	C	200	Physical limit=1 Logical limit=8 No	$425	$2,085
*4	64	D	250	Physical limit=1 Logical limit=6 No	$425	$2,510
(c)						
1	64	A	Logic change	Logical limit=6 Yes	$3,236	$3,236
*2	63	E	350	Physical limit=1 Logical limit=7 No	260	$3,496
*3	62	C	200	Physical limit=1 Logical limit=8 No	$425	$3,921
*Optimal time-cost relationship						

Figure 7.5 Highway bridge, least-cost expediting.

of these newly critical activities discloses that it is not possible or feasible to shorten any of them. Therefore, the total effect of step 1 in Figure 7.5*b* is a reduction of three days. Steps 2, 3, and 4 shown in Figure 7.5*b* reduce the highway bridge to a duration of 64 working days at a total additional cost of $2,510. The sum of $2,510 is the least additional cost for which the project can be shortened by six days.

To reduce the highway bridge duration below 64 days, a third new series of shortenings is needed. Step 1 of Figure 7.5*c* shows that prefabricating and using two sets of abutment forms (a logic change, not an activity shortening) reduces the highway bridge duration from 70 to 64 working days, a gain of six days. The time reduction achieved by a logic change is the difference between the lengths of the critical paths before and after the logic revision is made. A change in network logic, therefore, has no physical limit, only a logical limit. Steps 2 and 3 in Figure 7.5*c* reduce the duration of the highway bridge to 62 days at an additional cost of $3,921. It might be noted here that if activity 250 is now expedited, no further shortening of the project results. This is because when the logic change in step 1 of Figure 7.5*c* is made, there is a change in critical path location, and activity 250 is no longer critical. Expediting it will only increase its float, not reduce the length of the critical path.

The project duration has now been reduced by a little more than 10 percent, the original objective. Hence the time reduction study is now complete and project management must decide how much expediting it wishes to pay for, if any. Figure 7.6 summarizes the expediting costs involved in shortening the highway bridge. This information tells the contractor how to reduce the project duration by any given number of working days at the least cost, up to a maximum shortening of eight days.

7.18 LIMITATIONS ON TIME REDUCTION STEPS

As has been shown, a number of limitations apply to how much time reduction can be accomplished in any one step. These limitations are summarized in the following:

1. *Physical Limit of a Critical Activity.* This is the maximum shortening of a given activity that is considered to be practicable. Although most activities can be shortened to some extent, some are considered to be intractable on practical grounds.

2. *Logical Limit of a Critical Activity.* The reduction in duration of a critical activity reduces the total floats of other activities, this sometimes causing another chain of activities to become critical. This can, at times, prevent expediting an activity to its full potential. Step 1 in Figure 7.5*b* is an example of this.

3. *Logical Limit of a Network Logic Change.* A network logic change results in a set number of days being gained with no time possibilities in between. Step 1 in Figure 7.5*c* is a logic change that reduces the highway bridge duration by six days. The six days is an irreducible time reduction, it is that or nothing.

4. *Shortening Limited by a Parallel Critical Path.* Parallel critical paths or subpaths are common. If one branch of parallel critical paths is to be decreased

in length, a commensurate decrease must also be made in the other branch. If only one branch is reduced, this simply makes it a "floater," and neither the remaining critical path nor the project is reduced in duration. This situation was encountered in conjunction with the inability to completely shorten activity 60 in Section 7.17.

5. *Shortening Limited by an Irreducible Critical Path.* When any given critical path has been shortened to its full capability, no further reduction in project duration is possible. A critical path that cannot be compressed makes fruitless the shortening of nonassociated portions of the network.

7.19 VARIATION OF TOTAL PROJECT COST WITH TIME

Prior discussions have shown that a general characteristic of construction projects is for their direct costs to increase and their field overhead costs to decrease as the construction period is reduced below the normal time. Figure 7.6 is a plot of the increase in direct cost required to expedite the highway bridge. It was earlier determined in Section 7.7 that the time variable field overhead expense for this project amounts to approximately $367 per working day. Figure 3.7 disclosed that the project constant overhead expenses totalled $15,548.00. Figure 7.7 shows how total project overhead, direct cost, and the sum of the two vary with time. The field overhead expense shown in Figure 7.7a is obtained by multiplying the number of working days by $367 and adding the sum of $15,548.00. The normal direct cost of the project is procured from Figure 3.8 as $216,814. Adding the expediting costs from Figure 7.6 to this project normal cost gives the direct costs shown in Figure 7.7b. Combining the direct cost with the overhead expense for each project duration gives the costs shown in Figure 7.7c. Total project cost for any duration may be obtained by adding small tools ($2,809.00), tax ($7,825.38), and cost of the contract bonds ($3,752.16) to the values shown in Figure 7.7c. The three values just sited are obtained form the original project estimate given in Figure 3.8.

To illustrate how the plot in Figure 7.7c can be used, suppose that the construction period prescribed by contract for the highway bridge is 62 working days and that liquidated damages in the amount of $300 per calendar day will be imposed for late completion. The best evidence now available indicates that 70 working days will actually be required unless the job is expedited. If a management decision is made not to expedite, the contractor is apt to be assessed $3,000 in liquidated damages (eight working days = 10 calendar days @ $300). If the job is expedited to a duration of 62 working days, Figure 7.7c indicates that the total additional expense to the contractor will be $259,022 – $258,037 = $985. Very likely the decision of project management will be to expedite the project.

There are reasons other than project shortening why a contractor must be able to associate costs with project duration. One such instance is when the contractor is bidding a project and the owner requires that the contractor must quote separate prices for different specified construction periods. Although this requires contractor action before a complete time study is usually available, the contractor must still make some judgments regarding how direct and overhead costs vary with time.

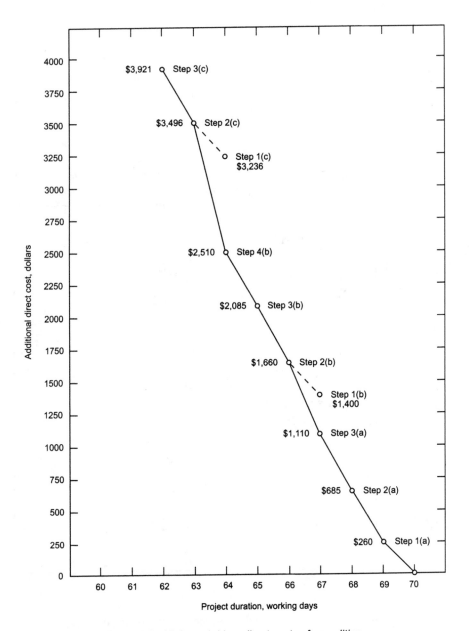

Figure 7.6 *Highway bridge, direct costs of expediting.*

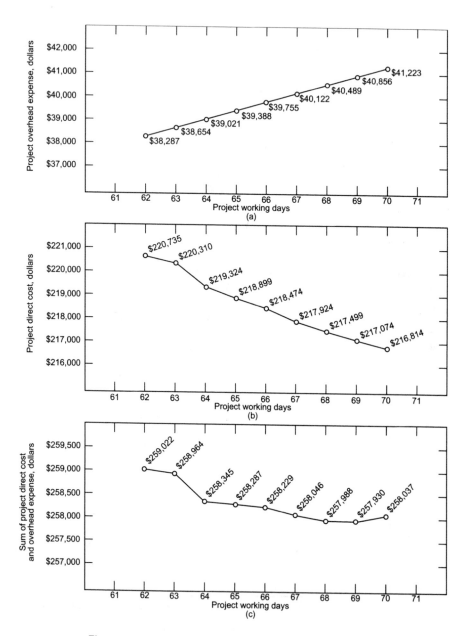

Figure 7.7 *Highway bridge, variation of costs with duration.*

7.20   EXPEDITED HIGHWAY BRIDGE SCHEDULE

Assume that the contractor has decided to expedite the highway bridge by eight working days to an anticipated duration of 62 days. This expediting information must now be reflected in the construction schedule. The results of the network recalculation made after step 3 in Figure 7.5c will provide this information.

Reference to Figure 7.5c shows that the following changes must be made to the original precedence diagram, Figure 5.1, when the recomputations are made.

1. The logic change accomplished by building two sets of abutment forms and starting activity 240 immediately after activity 230 is accomplished by eliminating the dependency line from activity 220 to activity 240 (Step 1 of Figure 7.5c).
2. Because two sets of abutment forms are now going to be used, the duration of activity 80, "Make abut. forms," is increased from three to six days (Step 1 of Figure 7.5c).
3. The duration of activity 350 is reduced from three days to two days (Step 2 of Figure 7.5c).
4. The duration of activity 200 is reduced from two days to one day (Step 3 of Figure 7.5c).

The four network changes just described are made to the precedence diagram of the highway bridge and the recalculations performed. Figure 7.8 summarizes the activity times and float values that are obtained.

7.21   MILESTONE AND INTERFACE EVENTS

The principal point of the preceding discussions has been the reduction of overall project duration. This includes the expediting of an ongoing project that has suffered previous delays. It has already been pointed out, however, that it is sometimes necessary to advance the expected dates of milestone and interface events. If the scheduled time of such an event does not satisfy an established time requirement, then some action by the contractor is in order. The procedure is very much the same as that for shortening project duration, except that the project critical path may or may not be involved.

The early time for any event is determined by the longest path from project start to that event, and efforts to advance the event time must be concerned with shortening this longest path. The longest path would first be restudied to see if the desired shortening could be achieved at no increase of direct cost. If not, then the contractor must resort to expediting at additional direct expense.

Activity (Bold type denotes critical activity) (1)	Activity Number (2)	Duration (Working Days) (3)	Earliest		Latest		Float	
			Start (ES) (4)	Finish (EF) (5)	Start (LS) (6)	Finish (LF) (7)	Total (TF) (8)	Free (FF) (9)
Start	0	0	0	0	0	0	0	0
Prepare & approve S/D abutment & deck rebar	**10**	**10**	**0**	**10**	**0**	**10**	**0**	**0**
Prepare & approve S/D footing rebar	20	5	0	5	7	12	7	0
Order & deliver piles	30	15	0	15	1	16	1	0
Move in	40	3	0	3	10	13	10	0
Prepare & approve S/D girders	50	10	0	10	1	11	1	0
Fabricate & deliver abutment & deck rebar	**60**	**15**	**10**	**25**	**10**	**25**	**0**	**0**
Fabricate & deliver footing rebar	70	7	5	12	12	19	7	6
Prefabricate abutment forms	80	6	3	9	19	25	16	16
Excavate abutment #1	90	3	3	6	13	16	10	0
Mobilize pile driving rig	100	2	3	5	14	16	11	10
Drive piles abutment #1	110	3	15	18	16	19	1	0
Excavate abutment #2	120	2	6	8	18	20	12	10
Forms & rebar footing #1	130	2	18	20	19	21	1	0
Drive piles abutment #2	140	3	18	21	20	23	2	0
Pour footing #1	150	1	20	21	21	22	1	0
Demobilize pile driving rig	160	1	21	22	24	25	3	3
Strip footing #1	170	1	21	22	22	23	1	0
Forms & rebar abutment #1	**180**	**4**	**25**	**29**	**25**	**29**	**0**	**0**
Forms & rebar footing #2	190	2	22	24	23	25	1	0

Figure 7.8 Highway bridge, expedited schedule.

Activity								
Pour abutment #1	200	1	29	30	29	30	0	0
Pour footing #2	210	1	24	25	25	26	1	0
Strip & cure abutment #1	220	3	30	33	30	33	0	0
Strip footing #2	230	1	25	26	26	27	1	0
Forms & rebar abutment #2	240	4	26	30	27	31	1	0
Pour abutment #2	250	2	30	32	31	33	1	0
Fabricate & deliver girders	260	25	10	35	11	36	1	1
Rub concrete abutment #1	270	3	33	36	44	47	11	11
Backfill abutment #1	280	3	33	36	33	36	0	0
Strip & cure abutment #2	290	3	32	35	33	36	1	0
Rub concrete abutment #2	300	3	35	38	44	47	9	9
Backfill abutment #2	310	3	35	38	39	42	4	4
Set girders	320	2	36	38	36	38	0	0
Deck forms & rebar	330	4	38	42	38	42	0	0
Pour & cure deck	340	3	42	45	42	45	0	0
Strip deck	350	2	45	47	45	47	0	0
Guardrails	360	3	47	50	49	52	2	2
Paint	370	5	47	52	47	52	0	0
Saw joints	380	1	45	46	51	52	6	6
Cleanup	390	3	52	55	52	55	0	0
Final inspection	400	1	55	56	55	56	0	0
Contingency	410	6	56	62	56	62	0	0
Finish	420	0	62	62	62	62	0	0

7.22 PROJECT EXTENSION

The entire emphasis of this chapter has been directed toward the shortening of a project. It may also be desirable, at times, to lengthen certain activities or even the project itself. An example of this might be a project whose costs were originally estimated assuming that expediting actions would be needed if the owner's time requirements were to be met. It is not unusual for a contractor to anticipate time difficulties and build into the original cost estimate the extra expense of overtime, multiple shifts, and other means of expediting the work. Unfortunately, at the time the project is being bid, the contractor will not usually have made an accurate forecast of project duration nor have identified the critical activities. About the only way that the contractor can figure in the extra costs of expediting actions is either to expedite most or all of the job operations or make some educated guesses about which work items may turn out to be critical.

As sometimes happens, a subsequent detailed network analysis reveals that all or some of the planned expediting procedures will not be necessary. Here is a case just the opposite of that previously treated in this chapter. The contractor now finds it desirable to relax the job plan and realize the attendant savings. Obviously, the contractor wants to do this in a manner that will maximize its gain.

Even if the overall project duration must remain as planned, the contractor can usually rescind the expediting actions for at least some of the noncritical activities. The duration increase allowed any given activity would have to be limited to its total float, a relaxation that would result in a new critical path or subpath. This means that a record must again be maintained concerning the effect of each activity's time change on the floats of other activities. If all expediting actions of noncritical activities cannot be rescinded, then the most expensive of these should be eliminated first.

If the overall project duration can be extended from the original plan, then certain of the critical activities can also be relaxed. The only effect of this is to lengthen the critical path and, hence, to increase the floats of all noncritical activities. In this case, the critical activities should be relaxed first, beginning with those most expensive to expedite. After this, the noncritical activities can be treated as before.

8

RESOURCE MANAGEMENT

8.1 OBJECTIVE

The completion of a construction project at maximum efficiency of time and cost requires the judicious scheduling and allocation of available resources. Manpower, equipment, and materials are important project resources that require close management attention. The supply and availability of these resources can seldom be taken for granted because of seasonal shortages, labor disputes, equipment breakdowns, competing demands, delayed deliveries, and a host of associated uncertainties. Nevertheless, if time schedules and cost budgets are to be met, the work must be supplied with the necessary workers, equipment, and materials when and as they are needed on the job site. This chapter discusses methods and procedures involved with the management of these three resources. Money, another project resource that requires close management control during the construction process, is discussed in Chapter 11.

The basic objective of resource management is to supply and support the field operations so that established time objectives can be met and costs can be kept within the construction budget. Field supervisors can achieve favorable production rates and get the most from their workers and equipment only when the requisite ways and means are optimally available. It is the responsibility of the project manager to identify and schedule future job needs so that the most efficient employment is made of the resources available. The project manager must determine long-range resource requirements for general planning and short-term resources for detailed planning. He must establish which resources will be needed, when they must be on site, and the quantities required. Arrangements must be made for their timely arrival with regular follow-up actions taken to ensure that promised delivery dates are kept. Where shortages, conflicting demands, or delays occur, the project manager must devise appropriate remedial measures. The project plan and schedule may have to be modified to accommodate or work around supply problems.

The scheduling and allocation of workers, equipment, and materials are interrelated. An action affecting one often affects the others in some manner.

8.2 PROJECT RESOURCE MANAGEMENT

With respect to resource management on construction projects, a few general observations at this stage can serve as valuable guidelines for the practitioner. The long-term leveling of resources provides a good indicator of future resource needs but only from a general planning point of view. Detailed planning months into the future is unnecessary and usually a waste of time. Detailed resource leveling has its major advantage when applied to the near future; that is, a maximum of 30 calendar days. Ample float makes efficient resource management possible while low float values almost inevitably mean schedule delays or the need for additional resources. The concept of float providing resource efficiency becomes important in cases of dispute with the owner regarding the "ownership" of float. A common case of this is where the owner orders extra work to be done but refuses to grant additional time to the contractor because the added work is not on the critical path.

The highway bridge is used to illustrate resource management procedures throughout this chapter. In actuality, the resource planning and scheduling for the highway bridge would almost certainly not be done independently of the several other parts of the Example Project or even of other company projects. The obvious reason for this is that there are usually conflicting demands for the same limited resources from other job sites. There will surely be instances where workers, equipment, and materials will have to be traded back and forth among the various Example Project segments to achieve maximum use of the resources available.

Resources must often be allocated on a project-wide or company-wide basis, with some system of priority being established among the separate sources of demand. This is a complex and difficult problem for which only partially satisfactory solutions are possible. Management action in this regard cannot be stereotyped, but must be based on judgment and economic factors intrinsic and unique to the particular situation. To develop the basic principles of resource management, the ensuing discussion treats the highway bridge job as if it were a totally separate and autonomous unit. These same principles will provide management guidance where multiconstruction sites are involved.

8.3 ASPECTS OF RESOURCE MANAGEMENT

Basic to manpower management is a detailed compilation of daily labor requirements, by craft or crew, needed to maintain the established project schedule. If it appears that there will be adequate numbers of workers available to satisfy these projected requirements, then the work presumably can be staffed sufficiently to maintain the established schedule, and no adjustment of the job completion date is likely to be required. However, peaks and valleys in anticipated daily labor demands are usual and some leveling or smoothing may be in order.

If the labor requirement takeoff discloses that the demand will exceed the supply, the management situation can become considerably more complex. Remedial measures to combat an inadequate labor supply can include diverting labor from noncritical to critical activities or resorting to some method of expediting the critical activities. The "stretch-out" of noncritical activities or the use of overtime or subcontracting on critical activities may make it possible to maintain the originally established schedule. Otherwise, the project manager faces the difficult task of allocating the available labor to the various activities in such a way that it will be used to best advantage and minimize the project time overrun.

With respect to equipment management, most of the major decisions concerning how the job is to be equipped were made at the time the job cost was estimated. Nevertheless, it is the responsibility of the project manager to see that the job is properly and adequately equipped. In a manner analogous to the checking of labor needs, a compilation is made of the total equipment demand on a daily basis. If there are conflicts among project activities for the same equipment items, rescheduling of noncritical activities will often solve the problem. Lacking this, working overtime, on weekends, or multiple shifts might circumvent the difficulty. Another solution could be to arrange for additional items of equipment to be supplied from some outside source. If excessive equipment demands cannot be ameliorated in one of these ways, the conflict must be removed by rescheduling activities with the least possible increase in project duration.

Material management on a construction project is essentially a matter of logistic support. Job materials in the proper quantity and specified quality must be available at the right place and time. All aspects of material procurement, from ordering to delivery, are directed toward this objective, and a positive system of checks and controls must be established to assure its realization.

Subcontractors can and often do play an important role in achieving project time and cost goals. Although there is much variation in actual practice, subcontractors typically provide their own workers, equipment, and materials. The project manager will seldom have any direct voice in the management of subcontractor resources. Rather, the manager's responsibility here is more one of ensuring that each subcontractor commences work at the proper time and processes the work in accordance with the established time schedule.

8.4 TABULATION OF LABOR REQUIREMENTS

The management of construction workers begins with a tabulation of labor needs, by craft, for each project activity. Figure 8.1 is a take off of the general contractor's labor requirements for the highway bridge. This figure does not include the manpower needed by the reinforcing steel and painting subcontractors. Much of the information contained in Figure 8.1 is readily available from the original estimate in those instances where labor cost was estimated using crew size and production rate and the activity involves only one cost account. For instance, Figure 3.6 shows one foreman, one cement mason, six laborers, one equipment operator, one oiler, and one carpenter as the crew to pour the abutment concrete (activities 200 and 250). These labor requirements are entered directly into Figure 8.1. In doing so, it is assumed that the crew foreman will be a carpenter and, consequently, a requirement for two carpenters is entered for activities 200 and 250. Similar assumptions regarding craft foremen are used throughout Figure 8.1. Where the labor cost was initially estimated by using a unit cost, crew sizes assumed when activity durations were being estimated can be used (see Section 5.4).

If activities involve more than one cost account and, correspondingly, more than one crew, the determination of their labor requirements is less direct. To illustrate this point, activities 90 and 120 both involve abutment excavation. As indicated by Bid Item 1 and Bid Item 2 of Appendix A, each of these activities involves two different cost accounts: excavation, unclassified and excavation, structural. It was assumed in the planning stage of the highway bridge that the unclassified excavation would be performed with a tractor-dozer and would be done for both abutments in a single operation. Once this work is completed, the structural excavation will be done first for abutment #1 and then for abutment #2. Thus activity 90 is really the unclassified excavation for both abutments as well as the structural excavation for abutment #2. This is reflected in the original plan with a three-day duration for activity 90 and two days for activity 120. This is the rationale for the workers listed for these two activities in Figure 8.1.

8.5   PROJECT LABOR SUMMARY

With the information given in Figure 8.1 and the time-scaled network shown in Figure 5.14, it is an easy matter to determine the projected daily labor needs for the highway bridge based on an early start schedule of operations. Figure 8.2 is this project labor summary. Using the available floats of noncritical activities, it is possible to smooth or level the peak demands for manpower revealed by Figure 8.2. (Methods for doing this are discussed later in this chapter.) The early start labor requirements usually serve as the best starting point for any adjustment that may be required of the daily labor demands. However, it must be noted at this point that an early start schedule often turns out to be inefficient in terms of cost and resources.

An important characteristic of the highway bridge is revealed by the labor compilation in Figure 8.2. If the early start schedule is followed, there will subsequently be a period of time during which no work can proceed because of the lack of materials. After the abutment forms are built and the excavation is completed, the job will be at a standstill for seven working days awaiting delivery of piles and reinforcing steel. This is clearly disclosed in Figure 8.2 by the seven-day gap in the project labor requirement after the first few job operations have been finished. The

Activity Number	Activity	Duration	Pile-driverman	Carpenter	Laborer	Equipment operator	Oiler	Iron worker	Cement mason	Truck driver
40	Move in	3		3	4	2				4
80	Prefabricate abutment forms	3		3	3					
90	Excavate abutment #1 — days 1 & 2	3			3	1	1			
	— day 3				7	1	1			2
100	Mobilize pile driving rig	2	3		2	2	1			
110	Drive piles, abutment #1	3	4	1	2	1	1			
120	Excavate abutment #2 — day 1	2			3	1	1			
	— day 2				7	1	1			
130	Forms & rebar, footing #1	2		1	1					
140	Drive piles, abutment #2	3	4	1	2	1	1			
150	Pour footing #1	1		2	4	1	1		1	
160	Demobilize pile driving rig	1	3		2	2	1			2
170	Strip footing #1	1			2					
180	**Forms & rebar, abutment #1**	4		3	3					
190	Forms & rebar, footing #2	2		1	1					
200	**Pour abutment #1**	2		2	6	1	1		1	
210	Pour footing #2	1		2	4	1	1		1	
220	**Strip & cure, abutment #1** — day 1	3		3	6					
230	Strip footing #2	1			2					
240	**Forms & rebar, abutment #2**	4		3	3					
250	**Pour abutment #2**	2		2	6	1	1		1	
270	Rub concrete, abutment #1	3			1				2	
280	Backfill abutment #1	3			3	1				
290	**Strip & cure, abutment #2** — day 1	3		3	6					
300	Rub concrete, abutment #2	3			1				2	
310	Backfill abutment #2	3			3	1				
320	**Set girders**	2			4	1	1	4		
330	**Deck forms & rebar**	4		3	2			2		
340	**Pour & cure deck** — day 1	3		2	6	1	1		4	
350	**Strip deck**	3			3					
360	Guardrails	3			1		2			
370	Paint	5								
380	Saw joints — days 1 & 2	1			1					
390	**Cleanup** — day 3	3			7	1	1			2
					4					

(Bold denotes critical activities)

Figure 8.1   Highway bridge, activity manpower needs.

Working Days

Craft	1	2	3	4	5	6	7	8	9	10	11	12	13	14	15	16	17	18	19	20	21	22	23	24	25	26	27	28	29	30
Date	14	15	16	17	18	21	22	23	24	25	28	29	30	1	2	6	7	8	9	12	13	14	15	16	19	20	21	22	23	26
Piledrivermen		3	3	3	3											4	4	4	4	4	4	3								
Carpenters	3	3	4	3	3	3	3									1	1	1	2	2	3	1	1	1	2	3	3	3	3	2
Laborers	4	4	4	8	8	10	3	7								2	2	2	3	3	6	4	1	1	6	8	6	6	6	8
Equipment operators	2	2	2	3	1	1	1									1	1	1	1	1	2	2	1	1	1					
Oilers							1	1								1	1	1	1	1	2	1	1							
Iron workers																					1									
Cement masons																								1						
Truck drivers	4	4	4	2	2																									

June — July

Figure 8.2 Highway bridge, daily manpower compilation.

Craft	31	32	33	34	35	36	37	38	39	40	41	42	43	44	45	46	47	48	49	50	51	52	53	54	55	56	57	58	59	60	61	62	63
Piledrivermen																																	
Carpenters		2	3		3	3	3	3	2	2	3					3	3	3	3	2													
Laborers		8	9		7	7	7	3	6	6	6			8	8	6	2	2	2	6			4	3	3	1	1	1			7	7	4
Equipment operators		1							1					1	1					1													1
Oilers		1							1					1	1					1													1
Iron workers														4	4					4						2	2	2					
Cement masons		1			2	2	2		1					2	2	2											2	2					
Truck drivers																				4						1							1

	27	28	29	30	2	3	4	5	6	9	10	11	12	13	16	17	18	19	20	23	24	25	26	27	30	31	1	2	3	7	8	9	10
																			August											September			

Figure 8.2 continued.

job situation where an early start of field operations only results in a subsequent delay awaiting the receipt of key resources is a common one. This matter is discussed further in Section 9.4 with regard to preparing field schedules.

The labor summary of Figure 8.2 provides information about whether the local labor market can be expected to provide the numbers of tradesmen required. However, it is admittedly unusual for contractors to make such a labor takeoff purely to ascertain the adequacy of the labor supply. It is just assumed, for most projects, that the contractor will be able to hire sufficient workmen. The contractor's experience and knowledge of local conditions are valuable guides in this regard. However, this matter often deserves more attention than it gets. Severe shortages of certain labor skills often occur in many areas of the country during peak construction periods. Any arrangements to bring in workers with the requisite trade skills from outside areas must be made well in advance. Contractors will find that advance knowledge about the labor demands of their projects can be of considerable value in the overall planning and scheduling of their field operations.

8.6   VARIATION IN LABOR DEMAND

For present purposes of discussion, let it be assumed that there will be an adequate labor supply for the highway bridge, or at least that the peak demands can be smoothed off sufficiently to bring the demand within the supply. Figure 8.2 reveals that the requirements for different crafts vary widely and are, at times, discontinuous. Some variation in time demand for a given craft is normal because labor crews typically build up to strength at the start of the job and decline toward the end. However, very pronounced grouping of labor needs at different points during the construction period is decidedly undesirable and impractical. The recurrent hiring and laying off of personnel on a short-term basis is troublesome, inefficient, expensive, and scarcely conducive to attracting and keeping top workers. New tradesmen on the job are not as efficient as they are after they become familiar with the work involved. This is a "learning curve" phenomenon, where a crew's production goes up and its unit labor costs go down with task repetition. Then, too, there is the practical consideration that when workers are laid off for a few days, they may be difficult or impossible to replace.

8.7   MANPOWER LEVELING

Manpower leveling is the process of smoothing out daily labor demands. Perfection in this regard can never be attained, but the worst of the inequities can often be removed through a process of selective rescheduling of noncritical activities. On the highway bridge, the main crew is made up of carpenters and laborers. The process of manpower leveling discussed here will address only these two crafts. No amount of smoothing can even out the daily job requirements for equipment operators, ironworkers, cement masons, and pile driver operators. On the highway bridge, these workers will have to be provided when and as needed. It is for this reason that specialty crews are often shifted back and forth among company jobs. This kind of irregular labor demand often prompts general contractors to subcontract portions of their work.

The peak demands and discontinuities for carpenters and laborers on the highway bridge can be leveled to some extent by using the floats of noncritical activities. To illustrate the basic mechanism of how resource smoothing is accomplished through rescheduling, a simple example will be discussed. The reader is referred to Figure 8.3 which shows the daily requirements of the highway bridge for carpenters and laborers. These data are summarized in the form of two histograms to illustrate the peaks and valleys of labor demand required by an early start schedule.

Figure 8.3 has been prepared by plotting, for each activity, its daily demand for the designated crafts. Each activity is assumed to begin at its early start time. Opposite each activity and under the working days during which it will proceed is entered its daily labor demand. The symbol "C" is for carpenters and "L" for laborers. The labor totals indicated by the histograms are obtained by adding vertically the labor demand for each day. The information contained in Figure 8.3 assumes that the same sized crew will be used throughout the duration of any given activity. For activities that have long durations, this assumption is probably not justified. Generally, the crew for such an activity will start small, build up to full force, and taper off at the end. The assumption of constant crew size is reasonably accurate, however, for short duration activities such as are used on the highway bridge.

Figure 8.3 discloses that there is a peak requirement for ten laborers on project working day 6. This peak is caused by the fact that activity 80, requiring three laborers, and activity 90, requiring seven laborers, are both scheduled to be underway the same day. A usual way to remove or minimize such a conflict in an early start schedule such as Figure 8.3 is to move one of the conflicting activities to a later date. When a noncritical activity with no free float is moved to a later time to level a resource, any succeeding activities must be adjusted forward by a like amount. If free float is available, the finish of the rescheduled activity can be set later up to the amount of free float and not affect any following activity. Hence, when adequate free float is available, schedule changes to accomplish resource leveling are easily made. However, when a schedule adjustment bumps ahead a whole chain of succeeding activities, the resource needs of succeeding days can be affected. These changes may improve the overall situation or further complicate it. In the case of ten laborers on project working day 6 previously cited, reference to Figure 5.4 shows that activity 80 has 19 days of total float and 19 days of free float. Activity 90 has 12 days of total float and zero days of free float. This labor conflict can be easily remedied by moving activity 80 to a later date.

8.8   HEURISTIC MANPOWER LEVELING

A number of operations research techniques are available for obtaining optimal solutions to manpower leveling problems. Numerous algorithms are available to accomplish such a time critical analysis, but many of these require a computer to handle as few as two resources. Even on relatively small projects, these procedures can require large amounts of computer time. For this reason, heuristic methods are often used. Heuristic methods involve the application of approximations to solve very complex problems.

Approximations or rules of thumb are common in construction, and are often used to handle complex situations. Lacking precise information, as is often the case, such a rule provides prompt and reasonably accurate guidance. In the context of manpower leveling, heuristics can substantially simplify management decisions regarding the order in which activities should be rescheduled.

A simple manual heuristic method is used in Figure 8.4 to level the demand for carpenters and laborers on the highway bridge. The method presented is based on priority rules that give reasonable results when used with a modest number of labor resources. (In the case presented here, the number of resources is two: carpenters and laborers.) These rules are essentially the same ones that would be used intuitively by field supervisors if project management failed to provide them with a schedule of activity starts.

The priority rules referred to pertain to the order in which project activities will be rescheduled from an early start condition. In essence, the entire project is rescheduled. The critical activities are given highest priority and are scheduled first. The noncritical activities are then scheduled with priority being given to those activities with the earliest late start dates. When more than one activity has the same late start date, preference is given to the one with the least total float. The algorithm progresses through Figure 8.4 one day at a time, with activities being scheduled according to the priority rules just described. In this example, maximum limits on the daily labor demand have been set as three carpenters and seven laborers. These limits have been set equal to the maximum labor demands of individual critical activities as shown in Figure 8.1.

8.9 NUMERICAL EXAMPLE

In Figure 8.4, the basic information given in the left-hand columns is a list of activities on the highway bridge together with the resource need, duration, and late start value for each activity. The first step in the leveling process is to schedule the critical activities by listing the manpower required for each under the appropriate working days. Because critical activities 10 and 60 do not require labor, activity 180 is the first critical activity to be scheduled. It starts on day 26, finishes on day 29, and requires three carpenters and three laborers for each of its four days. Because the labor demand limitations are based on the maximum demands of the critical activities, all critical activities can be scheduled for the same dates that were called for in the original early start schedule. If the labor demand of a critical activity were to exceed the supply, it will not be possible to keep the project within its originally planned duration unless special measures are taken. This case is discussed in Section 8.11.

After all the critical activities in Figure 8.4 have been scheduled, noncritical activities are scheduled in order of their late start dates. The first of these is activity 40, which must start no later than day 12. This activity, along with all the other noncritical activities, is scheduled to start as early as possible subject to previously established resource limitations. Commensurately, it is scheduled to begin on day 1 and end on day 3. The activity with the next earliest late start is activity 90, with a late start of day 15. Because activity 90 follows activity 40 (see project network in Figure 5.1), activity 90 cannot be scheduled to start until day 4. When activity 90 is scheduled from day 4 through day 6, the project manpower limits are not exceeded, and its scheduling is acceptable. In a similar manner, activities 100, 110, 120, 130, and 140 are scheduled in order of their late start dates. Each activity is scheduled as close to its early start as possible but allowing the start date to slip as necessary to remain within the resource limits. It is to be noted that activity 80, which can start as early as day 4 based on the network logic, has been allowed to slip to day 14 in order to maintain resource levels. Once all of the activities preceding critical activity 180 have been scheduled, it is noted that there is a six-day gap when no resources are required. This is the same network characteristic previously noted in Section 8.5, where an early start of work on the site serves no purpose because of the delivery times required for pilings and reinforcing steel. As a result of this gap, all of the noncritical activities scheduled up to this point can be moved to the right in Figure 8.4 by at least 6 working days.

An important point must be made here. The logic of the network diagram shows that the start of field operations need not occur until day 12. The resource histograms in Figure 8.4 show that the start can be delayed six days without impacting resources. The point here is that float is necessary to allow for the shifting of activities for the efficient use of resources. Both the time consideration and the resource considerations are important. A decision regarding "Move in" will be discussed in more detail in Section 9.4.

A comparison of the labor demand histograms in Figures 8.3 and 8.4 shows that the leveling efforts have resulted in considerable improvement. For example, the peak requirement for laborers has dropped from ten to seven, and the number of days during which there is no need for laborers has been reduced from 15 to 12. However, if the opening activities were to be delayed by six working days as described previously, the number of days when there will be no need for laborers will be further reduced from twelve to six working days. The leveled labor demands shown in Figure 8.4 are obviously far from being uniform. In this regard, it must be mentioned that resource smoothing can be especially difficult on a project of limited extent with a large proportion of critical activities as is the case with the highway bridge. The total Example Project would provide a much more flexible basis for making significant and meaningful leveling studies.

8.10 MANPOWER LEVELING IN PRACTICE

The smoothing of labor demands from those shown in Figure 8.3 to those in Figure 8.4 is performed every day by field supervisors all over the country. The only difference is that the field supervisors perform this function in an intuitive and unordered way. Through long experience, good site supervisors have developed a talent and instinct for manning a construction job efficiently if not optimally. Manpower leveling, as a management procedure, is not standard practice in the construction industry. Rather, field supervisors are generally left to their own devices with regard to the field management of labor crews. This is not to say that such a procedure is necessarily bad. The ability to build up efficient and adequately sized crews and keep them busy is not only a usual attribute of skilled field supervision, it is also applied manpower leveling at its best.

Figures 8.3 and 8.4 are attempts to resolve an extremely complex matter with an enormously simplified job model. The data presented might be described as "theoretical" in a very untheoretical situation. The starts and finishes of activities are not nearly so definite as they are assumed to be in network usage, and there is usually a certain amount of activity overlap in practical fact. Neither is much of the job logic as rigid as it is made out to be. Tradesmen are shifted about from one activity to another as they are needed. The daily fluctuations shown in Figure 8.4 for laborers will not really occur during the construction period. A relatively stable laborer crew of five or six workers will be on hand throughout the job. A form of Parkinson's law takes effect, and every worker is kept busy, even when more laborers are present on a given day than Figure 8.4 indicates are actually needed.

Manpower leveling using present-day algorithms is a trial-and-error process and is made difficult by the fact that most activities use more than one labor classification as well as different types of equipment. A shift that improves one resource often complicates one or more other resources, either on that day or on succeeding days. Despite admitted shortcomings, however, manpower leveling can provide valuable information relative to the efficient use of job-site labor. Even if formal leveling studies are not attempted, the adopted rules of thumb pertaining to the priority of activity rescheduling can provide consistent and useful guidance. For example, it can be emphasized to field supervisors that critical and low float activities have manpower priority, and only those activities with large float values are to be used as "fill-ins." Another guide for field supervisors could be to save the available float on activities for possible later use in resource leveling. Where men and equipment are not needed elsewhere, activities should be started as early as possible, even those with large float values. On large and complex projects, or on jobs where the supply of labor is known to be limited, formal leveling efforts such as those previously discussed can provide valuable advance information for project management.

8.11 RESTRICTED MANPOWER SUPPLY

The preceding example of manpower leveling involved a case where only the rescheduling of noncritical activities was required to keep the job labor demands within established limits. Let it now be assumed that this condition is no longer true. This is a labor shortage situation which presents project management with a considerably more difficult rescheduling problem than that previously encountered. The basic implication of a labor shortage is that the durations of certain activities must inevitably be extended beyond their normal values if the manpower deficiency cannot be overcome by expediting actions such as subcontracting the work or working the available workers overtime. Subcontracting may well be the best answer, but such action depends on circumstances and will not be discussed further here. In any event, a restricted supply of manpower may or may not affect the overall project duration. In addition, recall that a time contingency is usually built into the scheduled project duration. One point is sure, however. If overtime or subcontracting cannot alleviate the situation, the contractor then has the problem of allocating the available labor among the activities to best advantage.

The first item to check when a craft shortage is expected is the requirement for this particular labor resource by the critical activities. Consider critical activity 330 on the highway bridge. This activity requires three carpenters per day for four days. Based on an 8-hour day, this would amount to 96 man-hours of carpentry work. If there are only two carpenters available, the contractor has a decision to make. If the activity is staffed by only two carpenters and if the usual 8-hour day is worked, the duration of the activity will be increased to six days. Because this is a critical activity, the project is commensurately delayed by two days. However, the project duration will be unaffected if the two carpenters work 12-hour days (four hours of overtime) while engaged on this activity. Some carpenter tenders (laborers) and perhaps other trades would probably also have to work overtime with the carpenters to keep the work in phase.

With regard to labor deficiencies on noncritical activities, the first action is to stretch out the duration of each activity concerned sufficiently to keep the labor requirement within the supply. However, when the extensions of noncritical activities equal or exceed the available floats, the contractor must again seriously consider the use of overtime. Otherwise, new critical paths and subpaths will materialize, possibly superseding the original critical path and delaying the entire project.

8.12 COMPLEX LABOR SCHEDULING

When labor shortages of several crafts are involved and overtime is not the answer, the matter of scheduling the available labor to minimize the project delay can become very complex and is referred to as a resource critical analysis. It is a trial-and-error process that involves allocating the available labor to the activities, all the while maintaining the necessary job logic to achieve project completion in the shortest possible time. In such cases, the complexity of the situation will normally preclude the obtaining of an optimal solution. The heuristic method described in Sections 8.8 and 8.9 works well with a restricted manpower supply. The critical activities are scheduled first as before. The durations of some critical activities will probably need to be extended or decisions made to expedite them. The noncritical activities are then scheduled using the same priorities as those applied previously; the earliest late start activities scheduled first with secondary priority given to those activities with the least float. When it is impossible to schedule an activity on or before its late start date because of manpower restraints, a new critical path has been formed. Activities that make up the new critical path must now be evaluated in the same manner as were the original critical activities and a decision made whether to expedite or delay the project. The best than can be done is to arrive at a practical compromise which appears feasible and reasonably efficient. At this stage, there are usually several critical paths through the project.

When deemed desirable and feasible, activities can be scheduled discontinuously at irregular time intervals. License to do this is, of course, dependent on the nature of the activity. Some job operations, such as a concrete pour, must be continuous. On the other hand, activities such as rubbing concrete or prefabricating concrete forms can be performed intermittently and serve well as fillers. In fact, this is probably how activity 270, "Rub concrete, abut. #1," and activity 300, "Rub concrete, abut. #2," of the highway bridge will be accomplished. This would be especially true for activity 300 because, as shown in Figure 5.14, if it is started at its early start time, it will parallel the setting of the steel girders. Performing both of these activities simultaneously could be unsafe and activity 300 can wait. Deck forming, pouring, and stripping are also possible safety conflicts with activity 300. Undoubtedly, the job superintendent will put the cement masons to work rubbing abutment #2 whenever an opportunity presents itself.

Much attention has been given elsewhere to the matter of complex resource scheduling. At present, such resource scheduling gives a good approximation of the total resource quantities needed and their approximate timing. It illustrates graphically that network float is often an illusion, being in fact the leeway necessary to schedule resources efficiently. When complex labor shortages occur that cannot be avoided by overtime or subcontracting, skilled and experienced field supervisors are probably better able to work out an acceptable daily schedule as the work goes along. It is the duty of the project manager to identify labor shortage problems well in advance, make the decisions necessary to minimize the problem, and allow sufficient time for the field superintendent to complete the job while allocating the available labor on a daily basis.

8.13 EQUIPMENT MANAGEMENT

On projects that require extensive spreads of construction equipment, how well the job schedule is maintained and production costs are controlled depend on the quality of equipment management in the field. The following considerations are important to the proper selection, use, and maintenance of equipment on the job.

1. To the maximum extent possible, equipment sent to the job should be of the type that will best perform the work under actual job conditions. Knowledgeable company personnel such as the project superintendent, equipment supervisor, and master mechanic should be consulted before final equipment selections are made. Equipment sizes should be matched to the production schedules, and equipment spreads should be balanced so that each unit delivers maximum production. Standardizing on brands helps to simplify maintenance and repairs.

2. Work should be planned and scheduled to achieve the fullest use of every equipment item. Idle equipment costs money. It is wise to make advance arrangements for replacement equipment when a unit is scheduled for repairs

or overhaul. Standby units of key equipment items, such as pumps, may save many times their cost.

3. Field maintenance should be a part of the prejob planning. It is difficult to overemphasize the importance of day-to-day maintenance to proper equipment operation. A systematic program of preventive maintenance is a vital part of field equipment management and compliance with established procedures should be checked. It is advantageous to keep on hand a supply of routine maintenance parts such as hoses, belts, filters, and so forth. The occasional cleaning of engines and cooling systems is an important aspect of machine care. Routine fueling and lubricating should be scheduled to minimize equipment downtime. A skilled and dependable person in charge of the equipment service truck is one of the most valuable people on the equipment team.

4. Repair services may be provided by the contractor's central equipment shop or by a nearby equipment dealer. What is important is that the facilities be capable of getting the machines back into production as quickly as possible. To attempt major repairs in the open under primitive conditions is poor practice. A good and conscientious mechanical supervisor in charge of repairs plays a key role in keeping equipment operating properly. Poor repairs mean breakdowns and lost production. Repairs and maintenance can often be scheduled for nights or weekends to minimize lost production time. Keeping a minimum inventory of commonly used repair parts and tire sizes can pay big dividends. When major repairs are scheduled, the necessary parts should be on hand before bringing the machine into the shop. It is good practice to clean and thoroughly check equipment before moving it from one job to another.

5. The production rate of an equipment unit depends on the operator and his field supervisors. Popular field supervisors will normally have a following of top equipment operators. Close supervision of operators and their equipment is as important as it is for a labor crew. A poorly functioning operator, besides not earning his wages, also impedes the production of an entire equipment spread. Some operators, attempting to get the fullest production, abuse their machines, causing breakdowns and repair costs. Constant supervision is required to achieve consistently good production rates.

6. There is sometimes a strong temptation to overload equipment in an effort to get more production. This is especially true with earth-hauling units. However, equipment is designed to do its best in the long run when loaded only to its rated capacity. The production gained by overloading will normally be more than offset by the additional costs of repairs and tires and the shorter life of the machine.

7. Actual production rates and costs should be checked on the site for each major piece of equipment. High repair costs can be indicative of worn-out machinery, inadequate equipment maintenance, or operator abuse. High unit costs of production can disclose improper equipment selection for the job, poor field supervision, improper functioning of the equipment, spread imbalance, or operator inefficiency.

8.14 EQUIPMENT SCHEDULING

To serve a variety of purposes, it is useful to have a daily resume of project equipment needs. Figure 8.5 is a compilation of the daily equipment requirements for the highway bridge based on early start dates. A time-scaled project network such as Figure 5.14 is very useful when information of this kind is being obtained. On larger and more extensive projects, it might be desirable to first make a tabulation of equipment needs by activity as was done for labor in Figure 8.1. Figure 8.5 will quickly disclose any conflicting demands among activities for the same equipment items. Actually, many of the duplicate demands for equipment are eliminated when the project logic is being developed. The planner instinctively detects and eliminates obvious overlapping needs for equipment as the job planning proceeds. Nevertheless, until Figure 8.5 or its equivalent has been developed, there is no formal way of checking for equipment conflicts.

When an equipment conflict does occur at this stage, it can be removed by rescheduling one of the activities involved by using its float or by adding a precedence constraint to the project network. Adding a precedence constraint means using a dependency line to show that one of the activities involved must follow the other rather than parallel it. The precedence arrow between activities 110 and 140 in Figure 5.1 is actually a resource constraint used to avoid conflicting demands for the crane and pile driving crew on the highway bridge. Figure 8.5 does reveal a conflicting requirement for the 50-ton crane on working days 21 and 31. The conflict of demand on day 21 will be used here to illustrate how this situation can be resolved. Figure 5.14 shows that activity 150, "Pour foot. #1," and activity 140, "Drive piles, abut. #2," both require the crane on that same day. This dual demand for the crane can be eliminated by moving activity 150 forward in time by one day. This activity has three days of float available. A precedence arrow from activity 140 to activity 150 would also remove the conflict by requiring that pile driving on abutment #2 be completed before pouring footing #1. A precedence arrow in the opposite direction would remove the equipment conflict, but would require the pile driving crew to delay three days between abutments #1 and #2 as well as having to rerig the crane for piledriving.

A word of warning must be expressed at this point regarding the use of precedence constraints to handle resource demands. The precedence constraint is a permanent solution to what may be a temporary problem. For instance, the precedence arrow from activity 140 to activity 150 would demand a strict logical relationship between them, even if activity 140 were delayed for a week or more. Resource leveling provides a form of soft logic that would not lead to the same condition.

If sufficient float is not available to permit activity rescheduling, the contractor will again be faced with extension of the project duration or project expediting. When the resource constraint concerns equipment rather than labor, the contractor has the additional alternatives of scheduling multiple shifts or bringing in additional equipment on a short-term basis. On projects where equipment is a major resource, it is advisable to go through an equipment leveling procedure as was described previously in this chapter for labor.

One of the real values of an equipment schedule is that it discloses gaps in the need for certain equipment items. For example, the early start schedule that forms the basis for Figure 8.5 shows that a loader will be required on the highway bridge job site at two different times, both for backfilling purposes. Large equipment is expensive to move on and off a job, and it is desirable to minimize such moves. Figure 5.4 reveals that activity 280, "Backfill abut. #1," has six working days of free float and thus can be deferred up to six days without affecting any other activity. When this is done, both abutments can be backfilled in immediate succession, and the loader need be moved to and from the job only once.

The equipment schedule in Figure 8.5 can be converted to a bar chart that shows the calendar dates during which various items of equipment will be required on the job. Such a schedule is useful, not only for the highway bridge job, but also as a guide for the overall allocation of company equipment to the pipeline relocation, the pipeline crossing, and other parts of the Example Project. Other company projects are thereby advised of equipment needs as well as equipment availability. If equipment must be brought in from other sources, the equipment schedule enables the project manager to give ample advance notice of project needs. If it is known that a piece of equipment will be idle for an appreciable period of time, and will not be needed on another part of the Example Project, the contractor may decide to rent it out or perhaps schedule the unit for repairs, maintenance, or overhaul. An approximate calendar schedule of project equipment needs is useful in a variety of ways.

8.15 COMPUTER APPLICATION

Perhaps more than in any other phase of construction project management, the computer has come to occupy an almost indispensable role in accomplishing effective resource leveling. The scheduling of numerous activities, each requiring its own combination of resources, becomes extremely complex and manual procedures quickly become practical impossibilities. Typical computer programs allow the user to assign as many as 50 separate resources to each activity. The distribution of each resource over the duration of the activity can be established to fit the situation. To illustrate, two units of resource A can be assigned to the first day of an activity, none on day two, and three units on the last day. This forms a resource requirement histogram for resource A for one network activity. Twenty resources distributed over 100 network activities, each with its own resource requirement histogram, give an idea of the complexity of resource leveling.

Computer programs also may allow the definition of a resource availability histogram for each project resource. In this way, for example, a crane can be scheduled to be available to the project from June 25 through August 30 and the computer will show this utilization during that period. In addition, the period of time during which one crane will not be sufficient will be pointed out. Some activities are defined to be interruptible; that is, they can be temporarily stopped and their resources used on another activity that has less float.

Equipment Item	Working Days (1–35)

Equipment Item	1	2	3	4	5	6	7	8	9	10	11	12	13	14	15	16	17	18	19	20	21	22	23	24	25	26	27	28	29	30	31	32	33	34	35
Dozer				1	1																														
Backhoe						1	1	1																											
50-ton crane	1	1	1	1	1											1	1	1	1	1	2	1	1	1		1	1	1	1	2					1
25-ton Hydraulic crane	1	1	1	1	1											1	1	1	1	1															
Pilehammer																1	1	1	1	1	1	1	1												
Compresor																1	1	1	1	1	1														
Flatbed truck		4	4	4																		1	1												
Lowboy				1	1																	1	1												
Front end loader																																			
Concrete saw					1																														1

	14	15	16	17	18	21	22	23	24	25	28	29	30	1	2	6	7	8	9	12	13	14	15	16	19	20	21	22	23	26	27	28	29	30	2
					June													July																	

Figure 8.5 *Highway bridge, daily equipment compilation.*

Equipment Item	36	37	38	39	40	41	42	43	44	45	46	47	48	49	50	51	52	53	54	55	56	57	58	59	60	61	62	63
Dozer																												
Backhoe																												
50-ton crane	1	1	1	1	1	1			1	1					1						1	1						1
25-ton Hydraulic crane																					1	1						
Pilehammer																												
Compresor																											1	
Flatbed truck																		1										
Lowboy	1	1							1	1		1																
Front end loader	1					1																						
Concrete saw																												

	1	3	4	5	6	9	10	11	12	13	16	17	18	19	20	23	24	25	26	27	30	31	1	2	3	7	8	9	10
									August														September						

Figure 8.5 continued.

All of these computer programs are heuristic in concept, and the algorithms can be extremely complex. Even so, they still produce only approximate results because they ignore many variables used intuitively by an experienced construction superintendent. The field supervisor will substitute a larger number of semiskilled laborers for a skilled workman, schedule selective overtime, or perhaps decide to prefabricate an assembly earlier for use at a later date when resources are in short supply. There are many options available in the field that cannot be identified in advance and programmed into a computer.

Computer programs often present the option of leveling resources with the number of resources fixed or the completion date established. With resources held to predetermined levels, the project duration is extended to stay within the resource limit. When time is held constant, the program predicts the minimum numbers of resources required. The predictions generated by the computer will not necessarily remain unchanged as the project advances. Even minor changes in activity durations or logic will cause the computer to completely reevaluate the project, moving activities back and forth to achieve a new set of levelled resources.

A resource leveling program is useful in evaluating the effect that the making of a change in the work has on the project time schedule. Changes in the work by the owner can sometimes have an effect on the critical path and result in a claim for extra time. This topic is discussed in Section 9.20. At this point, it is sufficient to say that changes to the project that reduce float values typically influence resource requirements. These impacts can be identified by holding the current number of resources constant, running the resource levelling program in the resource critical mode, and observing the amount of time the project has been delayed.

An important feature of resource levelling by computer is the ability to generate reasonably accurate resource predictions for the next 10 to 15 working days. This information enables the field managers to detect problems far enough in the future to cope with them. At the same time, the time frame is short enough that activity duration and logic changes will have minimal effect on resource levels.

8.16 MATERIAL SCHEDULING

Management control over materials is concerned with ensuring that the materials, in the quantities and qualities required, are on the job as needed. The contractor's purchase order customarily prescribes the quantity, quality, price, delivery date, and mode of transportation for the materials covered. The quantity and quality of all material deliveries are verified by inspection, count, and test (if necessary) as they arrive. Apart from these standard procedures, management control of job materials is directed primarily toward achieving their timely delivery. Considerable time and effort have been expended in developing a work schedule that will satisfy time and resource limitations. It should be obvious, however, that this schedule is meaningless unless it is supported by the favorable delivery of materials.

Lead times for material deliveries have already been included in the project schedule. This was accomplished by incorporating appropriate material restraints into the original project network. These restraints were based on the delivery terms included in the material quotations received from vendors when job cost was being estimated. They represent the times required for shop-drawing approval, material fabrication, and delivery to the job site. To the maximum extent possible, these lead times have been built into the operational schedule.

Immediately after the construction contract has been signed, it is necessary to fix the deadline dates by which purchase orders for the various project materials must be issued to the suppliers. In the case of the highway bridge, the critical path includes the preparation of shop drawings and the delivery of the abutment and deck reinforcing steel. The immediate issuance of this purchase order, the obtaining of the necessary shop drawing approvals, and the fabrication and delivery of this material are of major importance.

The latest possible order date for any particular item of material is set by noting when the work schedule requires the material to be on the job site and allowing for the delivery interval. The delivery period includes the times necessary for purchase order preparation and transmittal, shop-drawing preparation and approval, fabrication of the material, and delivery to the contractor. On the basis of this information, the project manager can prepare a purchase order schedule for the use of his company's procurement department. This schedule will contain all of the necessary purchasing information as well as the deadline date by which the order must be processed and transmitted to the vendor.

Establishment of the order lead time is an important aspect of material control. Ample provision must be made for the delivery interval, with some safety factor present to allow for unforeseen delays such as the required resubmittal of corrected shop drawings to the architect–engineer. Most field supervisors have a pronounced tendency to encourage the early ordering of materials because they regard this as insurance that the materials will be available when they are needed. Purchasing personnel are prone to "buy the job" completely soon after the contract has been signed. With price escalations being a possible problem, this is often a good policy. It is necessary, however, to coordinate material deliveries closely with the progress of the work. It is often undesirable to have excessively early delivery to the job site. Piles of unneeded materials can lead to serious problems of theft, damage, weather protection, and interference with job operations. There is the practical factor, too, that early material delivery by the vendor requires that the contractor expend funds perhaps better applied elsewhere at the time.

In the case of very restricted urban job sites, the careful scheduling of material deliveries is especially important. Early deliveries can be routed to temporary storage facilities, owned or rented by the contractor, when there is no suitable alternative. This does, however, involve additional expense of storage, handling, insurance, and drayage. When off-site material storage is resorted to, delivery to the job site must be anticipated in sufficient time to make the necessary arrangements. This particularly applies to congested downtown areas where permits, police escorts, and off-hours may be involved. Within limits, it usually pays to arrange for the delivery of materials to such job sites only shortly before they are needed.

8.17 SUBCONTRACTOR SCHEDULING

The management control of subcontractors centers about getting them on the job when they are needed and ensuring that they accomplish their work in accordance with the established job schedule. There are three main considerations involved in carrying out this responsibility. First, the project manager should consult with each of the major subcontractors during the planning and scheduling of the project. If the subcontractor has had an opportunity to participate in the preparation of the job schedule, it is instilled with a much greater appreciation of the role that it plays in the big picture and the importance of its performing as agreed.

The second consideration is the form and content of the subcontract agreement. A carefully written document with specific requirements in terms of submittals, approvals, and schedule can often strengthen the project manager's hand in obtaining subcontractor compliance. The timing of the issuance of subcontracts is not an issue here because a prime contractor will normally proceed with subcontract preparation immediately after the construction contract has been signed.

The third consideration is ensuring that the subcontractors order their major materials in ample time to meet the construction schedule. Some general contractors find it advisable to monitor their subcontractors' material purchases. This is sometimes accomplished by including a subcontract requirement that the subcontractor submit unpriced copies of its purchase orders to the general contractor within ten days after receipt of the subcontract. In this way, the general contractor can oversee the expediting of the subcontractors' materials along with its own.

In a manner similar to that for material ordering, the project manager must establish a lead-time schedule for notifying subcontractors when to report to the project. These notification dates are established on the basis of the project work schedule and a lead time that may vary from one week to a month or more. Subcontractors must be given adequate time to plan their work and move onto the site. Notification dates can be listed in chronological order of their appearance. As each notice date comes along, the project manager advises the subcontractor in writing of its report date and follows this up with a telephone call. In the interest of good subcontractor relations, the project manager should not schedule a subcontractor to appear on the site until the job is ready for it and its work can proceed unimpeded.

8.18 RESOURCE EXPEDITING

In a construction context as mentioned earlier, the term "expediting" can have two different meanings. Expediting, as it applies to project shortening, was discussed in Chapter 7. As used here, expediting refers to the actions taken to assure timely material delivery and subcontractor support for the project.

It is unfortunate that the stipulation of a required delivery date in a purchase order is no guarantee that the vendor will deliver on schedule. Neither does a letter of notification automatically ensure that a subcontractor will move in on the prescribed day. The project manager has an expediting responsibility to make sure that material and subcontract commitments are met. He may carry out the expediting actions himself or his company may have an expediting department. A full-time expeditor is sometimes required on a large project. On work where the owner is especially concerned with job completion, or where material delivery is particularly critical, the owner will sometimes participate with the contractor in cooperative expediting efforts.

A necessary adjunct to the expediting function is the maintaining of a check-off system or log where the many steps in the material delivery process are recorded. Starting with the issuance of the purchase order, a record needs to be kept of the dates of receipt of shop drawings, their submittal to the architect-engineer, receipt of approved copies, return of the approved drawings to the vendor, and delivery of the materials. Because shop drawings from subcontractors are submitted for approval through the general contractor, the check-off system will include materials being provided by the subcontractors. This is desirable because project delay occasioned by late material delivery is not influenced by who provides the material. This same documentation procedure is followed for samples, mill certificates, concrete-mix designs, and other submittal information required. General contractors sometimes find it necessary on critical material items to determine the manufacturer's production calendar, testing schedule if required, method of transportation to the site, and data concerning the carrier and shipment routing. This kind of information is especially helpful in working the production and transportation around strikes and other delays.

Each step in the approval, manufacture, and delivery process is recorded and the status of all materials is checked daily. At frequent intervals, a material status report is forwarded to the project manager for his information. This system enables job management to stay current on material supply and serves as an early warning device when slippages in delivery dates seem likely to occur.

The intensity with which the delivery status of materials and the progress of subcontractors are expedited depends on the float of the activities concerned. Critical activities, for obvious reasons, must be the most closely watched. If delays appear forthcoming for such activities, strong appropriate action is necessary. Letters, fax communications, telephone calls, and personal visits in that order may be required to keep progressing on schedule. Low float activities are about in the same category because little slippage in these can be tolerated before they, too, become critical. If priority among activities becomes necessary, this would obviously be established on the basis of ascending float values.

Weekly job meetings with all major suppliers and subcontractors can be very helpful. The project network, updated schedule, and latest material status report can be used as the basis for the agenda of these meetings. The advance recognition of impending trouble spots enables early corrective actions to be taken.

9

PROJECT TIME MANAGEMENT

9.1 THE TIME-MANAGEMENT SYSTEM

Up through its present stage of development, the project time-management system has concentrated on work planning and scheduling. An operational plan and a detailed calendar schedule have been prepared to meet project objectives. This includes any necessary project shortening and resource leveling that could be identified prior to the start of construction. The work can now proceed with the assurance that the entire job has been thoroughly studied and analyzed. To the maximum extent possible, trouble spots have been identified and corrective action has been taken to eliminate them. A plan and schedule have been devised that will provide specific guidance for the efficient and expeditious accomplishment of the work. Project management now shifts its attention to implementing the plan in the field and establishing a progress monitoring and information feedback procedure. The time-management system has entered the execution phase.

It is axiomatic that no plan can ever be infallible, nor can the planner possibly anticipate every future job circumstance and contingency. Problems arise every day that could not have been foreseen. Adverse weather, material delivery delays, labor disputes, equipment breakdowns, job accidents, change orders, and numerous other conditions can and do disrupt the original plan and schedule. Thus, after construction operations commence, there must be continual evaluation of field performance as compared with the established schedule. Considerable time and effort are required to check and analyze the time progress of the job and to take whatever action that may be required either to bring the work back on schedule or to modify the schedule to reflect changed job conditions. These actions constitute the monitoring and rescheduling phases of the time-management system and are the subjects of discussion in this chapter.

9.2   ASPECTS OF TIME MANAGEMENT

As a means of meeting the established time goals of a construction project in an environment of constant change, the field operations are subjected to periodic cycles of the time-management sequence shown in Figure 9.1. The basis for the time-management system is a current operational plan and schedule that are consonant with established project time constraints. Calculations based on the latest version of the job network produce a work schedule in terms of calendar dates for the start and finish of each project activity. This schedule is used for the day-to-day time control of the project. Such a system constitutes an effective early-warning device for detecting when and where the project may be falling behind schedule.

The work plan, however, must respond to changing conditions if project objectives are to be successfully accomplished. If the overall time schedule of a construction operation is to be met, there must be a way to detect time slippages promptly through an established system of progress feedback from the field. The monitoring

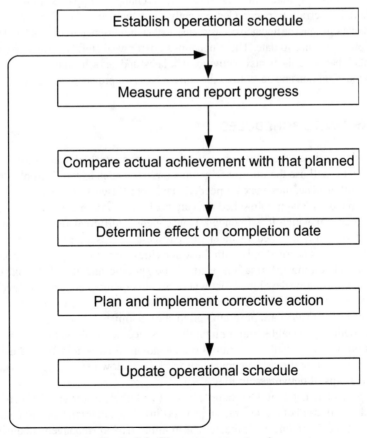

Figure 9.1   *Time management cycle.*

phase of time management involves the periodic measurement of actual job progress in the field and its comparison with the planned objectives. Project monitoring, therefore, involves the determination of work quantities put into place and the reporting of this information in a format suitable for its comparison with the programmed job schedule. Network activities constitute a useful and convenient basis for progress measurement and reporting.

At regular intervals, the stage of project advancement is observed and reported. As of each cutoff date, note is made of those activities that have been completed and the degree of completion of those activities that are in progress. Review of this information by project management discloses where the project is presently ahead or behind schedule and by how much. Critical activities and those with low float values are monitored very closely because of their strategic importance in keeping the project on schedule. Corrective action to expedite lagging work items is taken after analysis of the reported progress data reveals what options are available.

No project plan or schedule can ever be perfect and deviations will inevitably develop as the project progresses. As a result, the current version of the schedule will become increasingly inaccurate and unrealistic as changes, slippages, and other logic and schedule aberrations occur. Consequently, the network must be corrected as needed and calculations must be updated occasionally so that the current job schedule will reflect actual job experience to date. These updatings often reveal shifts in critical paths and substantial changes in the floats of activities. The latest updated schedule reflects the actual present job condition and constitutes the current basis for project time control.

9.3 KEY-DATE SCHEDULES

In formulating time schedules to be used for project control, consideration must be given to who will use the information. On a large job such as the Example Project, a hierarchy of schedules may be needed. The detail used in connection with the preparation of a given job schedule can be highly variable with the level of management for which it is intended. Consequently, different field schedules are prepared that are designed to meet the particular needs of the recipient. Craft supervisors are concerned only with those activities for which they are responsible, and their scheduling information needs to be specific and in substantial detail. Schedule information developed for owners, architect–engineers, and top-level field management will generally be in terms of milestones or key dates on which major segments of the project are programmed to start or finish.

One method of establishing a hierarchy of schedules is through the use of the project outline described in Section 4.6 and shown in Appendix B. The different levels of an outline provide a convenient work breakdown structure, consisting of various groups of network activities.

The project manager of the Example Project will likely be provided with a listing of key dates for each of the several subprojects involved. In total, this will constitute a master schedule of time goals that will be monitored by top management during the construction process. Figure 9.2 presents an initial key-date schedule that has been

		Scheduled		Required		Actual	
No.	Operation	Start	Finish	Start	Finish	Start	Finish
1	Procurement	Jun-14	Jun-19	Jun-14	Jun-19		
2	Field mobilization & site work	Jun-25	Jul-14	Jun-30	Jul-19		
3	Pile foundations	Jul-6	Jul-20	Jul-9	Jul-30		
4	Concrete abutments	Jul-20	Aug-17	Jul-20	Aug-30		
5	Deck	Aug-13	Aug-30	Aug-13	Sep-7		
6	Finishing operations	Aug-31	Sep-13	Aug-31	Sep-13		

KEY-DATE SCHEDULE

Project Highway Bridge

Project No. 9108-05

Figure 9.2 *Highway bridge, key-date schedule.*

prepared for the highway bridge. This key-date schedule is based on level two of the project outline. The original job plan and a 70 working-day duration constitute the basis for the presentation here. For reasons of clarity, the time shortening actions treated in Chapter 7 and the resource leveling of Chapter 8 are not carried forward to this discussion.

Early and late activity times discussed in Chapter 5 constitute the basis for most field schedules. Other names for these activity times are often used by contractors and by many standard computer programs. For example, early starts and finishes are commonly indicated as "scheduled" starts and finishes. Similarly, late starts and finishes are frequently shown as "required" starts and finishes. This new nomenclature is used with regard to the highway bridge schedules discussed in this chapter.

In Figure 9.2, the scheduled dates are the early start and finish times for the construction operations listed. Operations 1 and 2 are exceptions to this because of a job circumstance that will be discussed in the next section. The dates listed in Figure 9.2 have been obtained using the activity times listed in Figure 5.4 and converting these to calendar dates using the calendar in Figure 5.5. Figure 9.2 represents the first working version of a key-date schedule for the highway bridge and is prepared before field operations actually begin. Consequently, some of the dates contained in Figure 9.2 will probably change as the work progresses.

9.4 ADJUSTMENT OF MOVE-IN DATE

In making up the key-date schedule, it is noted in Figure 5.4 that activity 40, "Move in," has a total float of 12 working days. As depicted in Figure 5.14, activity 80, "Prefabricate abutment forms," activity 90, "Excavate abutment #1," and activity 100, "Mobilize piledriving rig," follow activity 40. Figure 5.4 shows that these activities have total floats of 19, 12, and 13, respectively. What all this means is that if these initial activities are scheduled and accomplished at their early times, field operations will then be at a standstill for several days waiting for the delivery of steel piles and reinforcing steel. Consequently, there is nothing to be gained by starting up field operations immediately. From a time standpoint, activity 40, "Move in," and the

184 PROJECT TIME MANAGEMENT

subsequent three activities could be delayed by as much as 12 working days without affecting the completion date. This delay would remove all of the float from these initial activities, resulting in a continuous chain of critical activities from "Move in" to "Cleanup." A delay of 12 days would leave no room for scheduling error and, as shown in section 8.9, would have an adverse impact on resource demands.

Based on both time and resource criteria, the project manager has decided to defer activity 40, "Move in," and the three following activities by nine working days, which will leave three days of float available for unanticipated problems. This postponement will in no way affect the project completion date and will eliminate an inefficient and unnecessary hiatus in field operations. Unfortunately, such a tardy move onto the site by the contractor can often cause considerable agitation to the owner unless the matter is explained beforehand. The ordering of reinforcing steel, girders, and steel pilings must still be done at the earliest possible moment, however.

9.5 DETAILED SCHEDULES

Considerably more detail than was included in Figure 9.2 is needed by the immediate work supervisors. The site superintendent on the highway bridge will require a substantially expanded time schedule that will provide a day-to-day forecast of field operations. Detailed project schedules are customarily prepared using activities as a basis. Different time information can be provided with reference to the activities in such schedules, but early start and finish dates are usual. Such a schedule is an optimistic one, and there are usually many instances where this schedule will not be met because of the inevitable delays and time slippages. Nevertheless, a project schedule based on activity early starts is the one generally used to establish project time objectives. The detailed job schedule should also note which activities are critical and indicate values of free float. This provides information of considerable significance to those who direct the work in the field. Labeling an activity as critical stresses its importance to lower-level managers. Knowledge of free float is also valuable in that it makes it possible for the field supervisor to use such extra time to meet unexpected job conditions.

The release of total float values to field supervisors is not always considered to be good practice. Total float, unlike free float, is usually shared by strings or groups of activities, and the use of such float in conjunction with one activity normally has an effect on the float values of other activities. Total float data can be misleading to those who are involved with only restricted portions of a project. Only the person responsible for the overall scheduling can evaluate and control the usage and allocation of total float. Free float is a readily usable commodity while total float must be carefully apportioned by knowledgeable planners.

On large projects, detailed schedules are prepared for each craft supervisor and subcontractor. Each of these schedules is a projected time program of the work for which that particular supervisor is responsible. The time spans of these schedules are limited, typically covering only the next two weeks to 30 days. This depends considerably on the nature of the work involved. It is seldom worthwhile to produce job schedules for more than a month in advance. There is no need for longer periods when much of such a schedule may be rendered obsolete by subsequent changes and updatings. Revised and updated schedules are issued as the work goes along. Tabular listings and computer printed bar charts are the most common forms of short-term work schedules. The network diagram is obviously a necessary adjunct to the job schedules. Copies of the network can be provided to the various supervisors or be kept in the field office.

The highway bridge is of such small size that only one detailed work schedule would be required, this for the job superintendent. The time schedule for the first month of construction operations on the highway bridge is shown in Figure 9.3. Move in and the three following activities have been delayed by nine working days as was explained previously in Section 9.4. As the work proceeds in the field, the job superintendent will enter the dates on which the activities actually start and finish into Figure 9.3. Work schedules, such as Figure 9.3, that are prepared for field supervisors may or may not include information concerning the delivery of job materials. Practice varies in this regard. This is not to say that such information is unimportant to the supervisor because this person is vitally concerned with what materials are ordered and when they will arrive. However, the scheduling of material deliveries is normally handled separately from the field operations. The project expeditor provides the project manager and site superintendent with periodic reports concerning the status of job materials, and this information is incorporated into the project weekly progress reports (See Section 9.10). Consequently, work schedules prepared for field supervisors often include only those activities that are physical parts of the work and omit material delivery information that is provided separately. This is the procedure followed herein and, hence, no material delivery information is included in Figure 9.3.

9.6 PROGRESS MEASUREMENT

To make periodic measurements of progress in the field, network activities serve as exceptionally convenient packages of work. The advancement of an activity in progress can be expressed in different ways. Commonly used methods are these:

1. Estimated number of working days required to complete the activity.
2. Estimated percentage completion of the activity in terms of time.
3. Quantities of work units put into place.

How a contractor chooses to express activity completion depends on the type of work involved and whether these same data are also used to check field costs. However, the number of working days required to finish an activity is fundamental to the workings of project time management. Progress data in the other forms are readily converted into days to complete by the following relationships:

$$\text{Working days to complete} = d\,(1 - P/100)$$

$$\text{Working days to complete} = d\,(1 - W/T)$$

DETAILED SCHEDULE

Project Highway Bridge

Project No. 9108-05

Prepared by G.A.S.

Date June 14

Activity	Activity Number	Scheduled Duration (Working days)	Scheduled Start Date	Scheduled Completion Date	Free Float (Working days)	Actual Start Date	Actual Completion Date
Move in	40	3	25-Jun	29-Jun	0		
Prefabricate abutment forms	80	3	30-Jun	2-Jul	10		
Excavate abutment #1	90	3	30-Jun	2-Jul	0		
Mobilize pile driving rig	100	2	30-Jun	1-Jul	1		
Excavate abutment #2	120	2	6-Jul	7-Jul	1		
Drive piles, abutment #1	110	3	6-Jul	8-Jul	0		
Forms & rebar, footing #1	130	2	9-Jul	12-Jul	0		
Drive piles, abutment #2	140	3	9-Jul	13-Jul	0		
Pour footing #1	150	1	13-Jul	13-Jul	0		
Demobilize pile driving rig	160	1	13-Jul	13-Jul	3		
Strip footing #1	170	1	14-Jul	14-Jul	0		
Forms & rebar, footing #2	190	2	15-Jul	16-Jul	0		
Pour footing #2	210	1	19-Jul	19-Jul	0		
Strip footing #2	230	1	20-Jul	20-Jul	8		
Forms & rebar, abutment #1	**240**	**4**	**20-Jul**	**23-Jul**	**0**		
Critical activities in bold							

Figure 9.3 Highway bridge, initial detailed schedule.

where

d = total activity duration in working days

P = estimated percentage of completion

W = number of work units put into place

T = total number of work units associated with the activity

Inherent in the above relationships is the assumption of a straight-line variation between time and work accomplishment. If an activity requires a total of four days for its performance, it is assumed that it is 25 percent complete after one day. This is normally an acceptable assumption if an activity is limited in scope. A more realistic relationship between time and work accomplishment may have to be used where activities are of substantial extent.

It is obvious that a time-management system is no better than the quality of the input information. If progress reports from the field are inaccurate, then management decisions will be made on the basis of fictitious situations. It is very important that progress measurements be done conscientiously and with reasonable accuracy. Management action must be based on what actually happened, not what should have happened. The person responsible for progress reporting must be impressed with the importance of factual and correct determinations. An important consideration here is that project progress records often are important in settling later disputes regarding project delays. Progress reports can form the basis for claims and litigation.

The conclusion of a given activity must be viewed in terms of its substantial completion rather than its absolute finalization. As work progresses in the field, there are many items of work which, at least temporarily, are not completely finished. Involved are small deficiencies that are subsequently remedied when the opportunity presents itself. Progress reporting must make allowances for such minor shortcomings.

9.7 PROGRESS REPORTING

How often field progress should be measured and evaluated depends on the degree of time control that is considered to be desirable and feasible for the particular work involved. Within limits, the greater the frequency of feedback and response, the more likely it is that the project time objectives will be met. However, this rule must be tempered with other considerations. For example, some kind of balance must be struck between the cost and effort involved and the management benefit gained. Another consideration is that the same field progress report often serves for field cost management (Chapter 10) as well as time management. Consequently, the cycle times for both management applications are frequently matched.

Fast-paced projects using multiple shifts may demand daily progress reports. On the other hand, large-scale jobs such as earth dams, which involve a limited range of work items, may use a reporting frequency of a monthly basis or even longer. It is difficult to generalize because management control must be consonant with project characteristics and peculiarities. On projects of the size, duration, and type of the highway bridge, progress reporting would probably be done on a weekly basis. This is the premise here.

The project manager must see that progress measurement and reporting are properly done and that the progress information receives prompt management review and analysis. A standard procedure for collecting and transmitting the weekly progress data must be established. Progress measurement requires direct visual observation in the field by someone familiar with the type of work involved. This may include a physical count of work units in place or may be reduced to evaluating quantities of work accomplished from the project drawings. At times, suitable measurement can come from delivery tickets for materials like concrete or load slips for earth moving. On many projects, the project manager personally carries out the measurement and reporting functions. Otherwise, a staff member, such as the field engineer, will perform these duties. In any event, an independent review of work accomplishment is to be preferred. All things considered, field supervisors are not usually the best choice for progress reporting. For one, they are very busy people who are not normally motivated toward paperwork of any kind. Additionally, a field supervisor may fail to report unfavorable progress in the hopes of working problems out later.

9.8 BAR CHARTS

On receipt of progress measurements from the field, the information must be compared with the latest project schedule. This can be done in different ways depending on management preferences and procedures. For example, a tabular listing can be prepared that shows the scheduled start and finish dates and the actual start and finish dates for each activity. Although progress data in this format may be useful for some purposes, such does not usually serve as the best medium for making a comprehensive evaluation of the current time status of construction operations.

For the day-to-day time management of a project, some form of graphic display is effective and convenient. A widely used method for recording job progress makes use of the bar chart, the general characteristics of which are discussed in Sections 5.28 and 5.29. There are several different styles and conventions used in drafting project bar charts. Two of the most common are presented herein. Both are widely used by the construction industry. One procedure is used to depict progress on the highway bridge as of July 14 and the other method is the basis for the July 21 progress report. For obvious reasons, the contractor would use the same bar-charting procedure throughout the construction period of the highway bridge. The two styles are mixed here so the workings of each can be observed.

The bar chart is an excellent medium for recording progress information and portraying the current time status of individual activities or other project segments. However, it is not a proper tool for evaluating the overall time status of the project or planning corrective measures when the work falls behind schedule. Only the project network can adequately perform this function.

Considerable time and expense can be involved with the manual preparation and updating of project bar charts. Computers are widely used to print out updated bar chart schedules. When project outlines are used for a work breakdown structure, bar charts of differing levels of detail can be produced with ease.

9.9 HIGHWAY BRIDGE AS OF JULY 14

Figure 9.4 has been prepared to show progress on the highway bridge up through the week ending Wednesday, July 14 (working day 22). This bar chart shows the scheduled and actual beginning and completion dates for each activity up through July 14. Plotted progress data have come from past field measurements that were made and reported at weekly intervals. Thus this bar chart is updated once a week and shows the current status of each activity and how its accomplishment compares with the schedule. Practice varies concerning the entry of information concerning material deliveries and other resource information on bar charts. In this book, the bar charts portray the advancement of the physical aspects of the work only and do not include dates for material deliveries or the availability of other resources.

In Figure 9.4, the upper row of shaded ovals for each activity extends between its scheduled start and finish dates. As has been discussed previously, bar charts are customarily made up on an early start basis. However, exceptions to this general rule on the highway bridge are the first four activities listed in Figure 9.4. As explained in Section 9.4, the scheduled beginnings of these four activities have been delayed by nine working days after their early start times. The white ovals that extend to the right of the noncritical activities represent the total float.

The lower row of black ovals opposite certain activities shows the time period during which work actually progressed on these activities. The start of each of these rows is plotted at the date on which work commenced. The line is then plotted to the right at weekly intervals, either to the current date or terminating at the day of completion. For each activity, the cumulative percentage completion is entered at the right end of the lower line until it is completed. The numbers 0 and 100 on the lower lines show the actual time periods required to go from 0 to 100 percent completion. If an activity is in process at the time of a progress measurement, but not yet completed, the actual percent completion is marked. An example of this is activity 140, which was only 30 percent completed on July 14, the date of the last progress report. Comparison of the actual and planned percentage completions as of a given date reveals the time status of an activity then in progress. One advantage of the form of progress recording used in Figure 9.4 is that it provides an exact historical record of calendar times when job activities were actually in process.

A review of Figure 9.4 will provide a quick, informative review of how the job stands timewise as of July 14. Activity 40, "Move in," was accomplished exactly as it had been scheduled. Activity 80, "Prefabricate abut. forms," took one working day longer than planned and finished six days late. However, it was completed within its free float period so there is no problem. The experience with activity 120, "Excavate abut. #2," was much the same. Progress on activity 140, "Drive piles, abut. #2," is far behind schedule. This activity should have been completed by July 13. Instead, it was only 30 percent finished as of July 14. Figure 9.4 shows that activities 160 and 170 should have been completed by the July 14 cutoff date, but they were not even started.

CONSTRUCTION

Project _____ Highway Bridge _____

Activity	Activity number		1	2	3	4	5		6	7	8	9	10
Move in	40	Scheduled											
		Actual										0	
Prefabricate abutment forms	80	Scheduled											
		Actual											
Excavate abutment #1	90	Scheduled											
		Actual											
Mobilize pile driving rig	100	Scheduled											
		Actual											
Excavate abutment #2	120	Scheduled											
		Actual											
Drive piles abutment #1	110	Scheduled											
		Actual											
Forms & rebar footing #1	130	Scheduled											
		Actual											
Drive piles, abutment #2	140	Scheduled											
		Actual											
Demobilize pile driving rig	160	Scheduled											
		Actual											
Pour footing #1	150	Scheduled											
		Actual											
Strip footing #1	170	Scheduled											
		Actual											
Forms & rebar, footing #2	190	Scheduled											
		Actual											
Pour footing #2	210	Scheduled											
		Actual											
Forms & rebar, abutment #1	**180**	**Scheduled**											
		Actual											
Strip footing #2	230	Scheduled											
		Actual											
Critical activities in bold			14	15	16	17	18	19	20	21	22	23	24	25	2

June

Figure 9.4   *Highway bridge, bar chart as of July 14.*

PROGRESS CHART

Date _____ July 14 _____

Job No. _____ 9108-05 _____

— Working Days ———————→

11	12	13	14	15			16	17	18	19		20	21	22	23	24		25	26	27	28	29		30

100

0 100

0 100

0 100

0 100

0 100

0 100

0 30

0 100

30

100

| 6 | 27 | 28 | 29 | 30 | 1 | 2 | 3 | 4 | 5 | 6 | 7 | 8 | 9 | 10 | 11 | 12 | 13 | 14 | 15 | 16 | 17 | 18 | 19 | 20 | 21 | 22 | 23 | 24 | 25 | 26 |

←——————————— July ———————————→

9.10   WEEKLY PROGRESS REPORTS

The reporting of work advancement in the field is accomplished by listing those activities that started, finished, or were in progress during the week just ended and indicating their stages of completion reached as of the cutoff date. Also noted are the dates on which activities either started or finished during the reporting period. These actual start and finish dates are entered by the field superintendent into the detailed field schedule, Figure 9.3, as the work progresses. There is no need to list activities previously completed or not yet started. Figure 9.5 is the weekly progress report as of Wednesday afternoon, July 21 (working day 27) on the highway bridge. Because there normally is only a limited number of activities in process during any given week, progress reporting is not usually a burdensome task, although some time is required if the measurements are properly determined. A file of the weekly progress reports can be maintained as a historical job record.

To report the overall project standing, the weekly progress reports must include procurement and delivery information as well as the measurements of physical progress. Each time that field progress is determined, a check is made with the job expeditor (see Section 8.18) to ascertain the delivery or availability status of materials, subcontractors, and construction equipment. Data pertaining to these resources are entered into the weekly progress report as shown in Figure 9.5.

Some explanation is required concerning the first two entries in Figure 9.5. These are the result of a change in delivery of the reinforcing steel for the abutments and bridge deck. The steel fabricator proposed an earlier rebar delivery for the abutments if the deck rebar could be delivered somewhat later. The abutment rebar (activity 60) was received on July 15 (originally promised July 19). A new activity 65, "Fabricate and deliver deck rebar," has been established with delivery promised August 9 (working day 40) or 13 working days from the date of the weekly progress report in Figure 9.5.

Progress reporting on a weekly basis is a common procedure in the construction industry and is being used herein for discussion purposes. This means that some day of the week must be chosen as the cutoff date. The day selected is often picked so that the same weekly progress measurement can serve for both time-management and labor cost-accounting purposes (Chapter 10). This is why Wednesdays are used here as progress-measurement days. Payday in the construction industry is frequently on Fridays and the daily time cards serve both payroll and labor cost-accounting purposes. When construction tradesmen receive their weekly wages on Friday afternoon, they are commonly paid up through the previous Wednesday. This gives contractors time to prepare and distribute their job payrolls. As a result of this, Wednesdays are frequently used by contractors as the cutoff days for weekly payrolls and, thus, their labor cost accounting as well. Having the field progress measurements on Wednesday afternoons thereby enables a single weekly measurement of field progress to serve for both time management and cost control. It must be understood that there are many exceptions to this procedure and that there is nothing sacred about using Wednesdays for progress-measurement purposes. The management principles discussed herein do not in any way depend on the day of the week used for cutoff purposes.

		WEEKLY PROGRESS REPORT			
Project Highway Bridge			Week ending	Wed., 21-Jul (Working day 27)	
Job No. 9108-05			Prepared by	GAS	
Activity	Activity Number	Date Started	Date Completed	Percent Complete	Working days to Complete
Fabricate & deliver abutment rebar	60	–	15-Jul	100	0
Fabricate & deliver deck rebar	65	–	–	–	13
Drive piles, abutment #2	140	12-Jul	–	80	2
Strip footing #1	170	15-Jul	15-Jul	100	0
Forms & rebar, abutment #1	180	16-Jul	21-Jul	100	0
Fabricate & deliver girders	260	–	–	–	10

Figure 9.5 *Highway bridge, weekly progress report.*

9.11 FIELD PROGRESS NARRATIVE

The weekly field progress report will usually be accompanied by a brief narrative discussion of salient project features. For example, it could include a general statement about the time status of the job, discussion of critical or low float activities now in difficulty or behind schedule, a description of potential trouble spots, and a noting of project areas that are going along exceptionally well. As a specific illustration of this point, Figure 9.4 shows that the pile driving for abutment #2 is in trouble and behind schedule. The weekly progress report that conveyed this information would explain that the difficulty was caused by the unsuspected presence of boulders and tightly packed gravel. An assessment of the problem would be given, including a forecast of how much delay would ultimately be involved.

9.12 JULY 21 STATUS OF HIGHWAY BRIDGE

The progress information contained in Figure 9.5 is now plotted on the project bar chart. Figure 9.6 is this bar chart updated through July 21. Recall that bar charts can differ about how actual progress is recorded. One procedure has been discussed in conjunction with Figure 9.4. Figure 9.6 depicts progress data in a completely different fashion.

In Figure 9.6, the early start schedule for each activity is plotted to a horizontal time scale as before. The schedule bar for an activity appears as a narrow horizontal white box extending from its scheduled start to finish date. The dashed line at the end of each activity represents that activity's total float. When using this second convention of recording actual work progress on a bar chart, it is assumed that the physical progress of an activity varies linearly with time. Actual progress, however, is not plotted to the established time scale as it was previously. Rather, the progress of an activity is indicated as a shaded portion of its schedule bar. The usual procedure is to shade in a length of the schedule bar in direct proportion to the physical advancement of the work. For example, activity 140 was reported by Figure 9.5 to be 80 percent complete as of the afternoon of July 21. In Figure 9.6, 80 percent of the length of the schedule bar for activity 140 has been shaded. The percent of completion

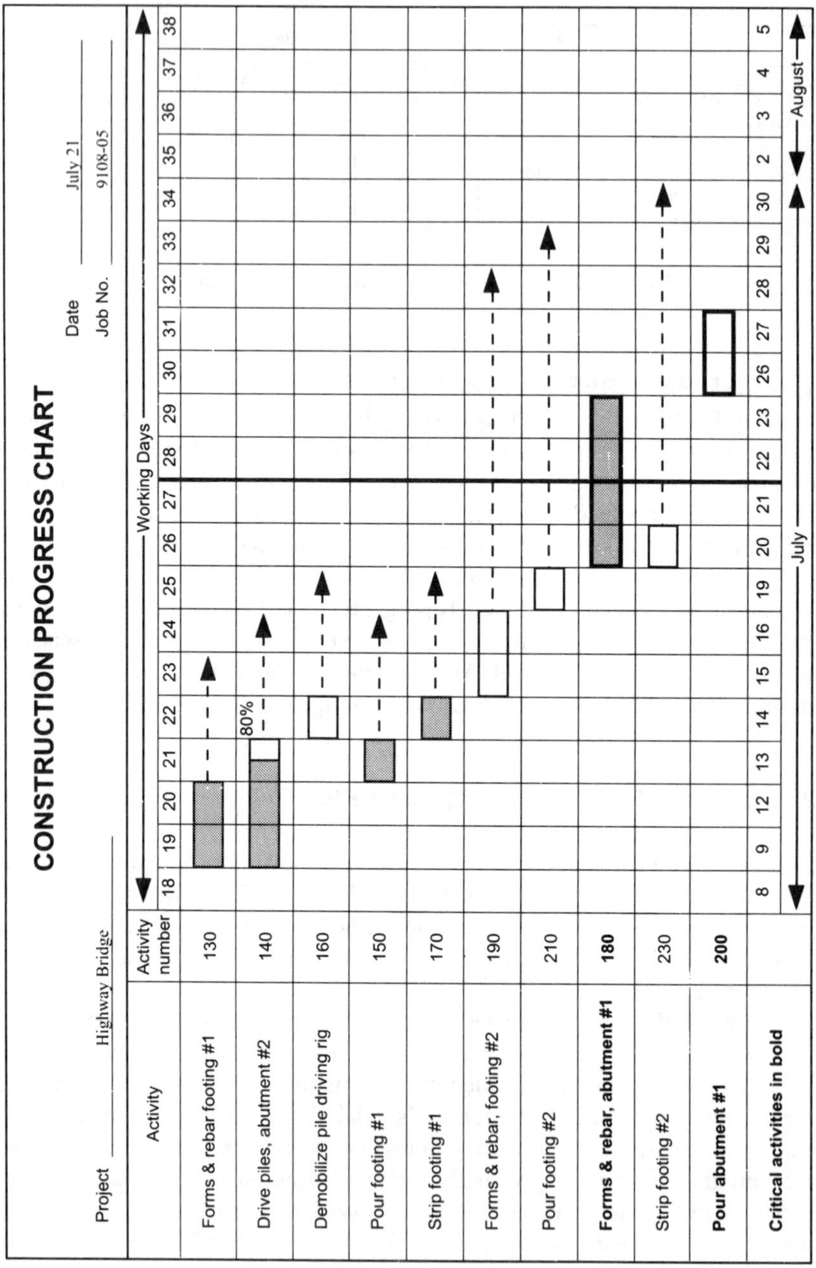

Figure 9.6 *Highway bridge, bar chart as of July 21.*

can also be entered above the bar if desired. If, at the time a progress report is made, an activity has been completed, the entire length of the bar is shaded.

The use of this mode of progress recording makes it possible to determine at a glance which activities were ahead and which were behind as of the cutoff date and by approximately how much. Figure 9.6 has been updated as of July 21 and a heavy vertical line has been drawn at this date. Any activity that has an unshaded bar to the left of the July 21 line was behind schedule as of that time. Activities 140, 160, 190, 210, and 230 are all in this category. Any activity that has a shaded bar to the right of the July 21 line was ahead of schedule. Activity 180 is an instance of this. The form of progress recording used in Figure 9.6 will not provide any historical record of true activity periods unless actual start and finish dates are jotted onto the bar chart as the work progresses.

Reference to Figure 9.6 shows that activity 140, "Drive piles, abut. #2," was only 80 percent completed as of July 21 and is now more than six working days behind schedule. This has prevented work from starting on footing #2. Figure 5.14 shows that activities 190, 210, and 230 have eight days of float associated with them. Whether this will be sufficient to absorb the delay in pile driving will be indicated by the next network update (see Section 9.15). As indicated by Figure 9.6, activity 180, "Forms and rebar, abut. #1," was finished two days ahead of schedule. This was made possible by the delivery of abutment rebar (activity 60) being two working days earlier than had been originally expected.

9.13 PROGRESS ANALYSIS

The analysis of job progress is concerned primarily with determining the effect of this latest information on the project completion date and any intermediate time goals that have been established. The successful attainment of set time objectives is, after all, the essential purpose of the time-management system. Considerable attention has been given to the fact that the length of the critical path determines the time required to reach a given network event. Practically, then, the analysis of progress data is concerned essentially with determining as closely as possible the present length and location of the applicable critical path.

When a progress report is received from the field, the status of the currently identified critical activities is probably the first item noted. This is a quick and simple check. If these critical activities have been accomplished by their scheduled finish times, there is no problem with the project time goal insofar as the current critical path is concerned. If a critical activity has been started late, a setback in project completion is likely unless the delay can be somehow made up by the time the activity is completed. If a critical activity has been finished late, the completion date is delayed commensurately.

The next step in the analysis is to check the possibility that a new critical path has been formed. This can be done by subjecting the noncritical activities to either of two different checks. One of these involves using either a "late start activity sort" or a "late finish activity sort." A late start sort is an activity listing in order of the latest allowable starting times (LS). A late finish sort is a similar listing in terms of the latest allowable finish dates (LF). These sorts can be obtained manually or by computer. If an activity has missed its LS date, the project is behind according to the present action plan. If an activity did not finish by its LF date, there is now a new critical path, and the job is delayed by the amount that activity completion trailed the LF time.

Another way to make much the same check involves the use of total float (TF) values. If the completion of an activity is delayed beyond its early finish (EF) time, its TF is reduced by the same amount. If the delay is equal to the TF, a new critical path is formed. If more than the TF is consumed, there is a new critical path of length longer than that of the original critical path and project completion is automatically set back by the amount of the overrun. The following discussion illustrates this point.

On the highway bridge as of July 21, activity 140, "Drive piles, abut. #2," is obviously behind schedule and it is now necessary to determine what effect this is having on the critical path location and job duration. Reference to Figure 5.4 shows that activity 140 has an LF value of 24. Figure 9.5, prepared as of working day 27, discloses that this activity is now expected to finish in 2 working days, or as of day 29. Consequently, activity 140 is now critical and there is a new critical path through the network. Noting in Figure 5.4 that activity 140 originally had a total float of 3 and learning that it will now finish 8 days behind its EF value of 21 provides the same information. The new critical path is 5 days longer than the previous one (64 working days), so project completion is now delayed by 5 days to a construction period (not including contingency) of 69 working days. Since the original project completion time was determined to be 70 days, including 6 days of contingency, there still remains one day of contingency available.

At this point it must be noted that the preceding discussion has been limited to the effect of only activity 140 on the duration of the highway bridge. The over-all time status of the project cannot be known until a complete updating calculation has been made as of the date of the last progress report (July 21). This will be done in Section 9.16.

9.14 CORRECTIVE ACTION

After each progress report has been analyzed, a decision must be made about what corrective action, if any, is required. Small delays, actual or potential, will not usually require any particular corrective action provided that the usual contingency allowance has been included in the project duration. With a contingency of only six working days included in the highway bridge schedule, the trouble with the driving of piles for abutment #2 will cause considerable concern. Where time slippages of the approximate length of the network time contingency become involved, then project management must give serious consideration to taking some form of corrective action. In general, remedial steps are indicated when situations like these arise:

1. Activities begin to fall appreciably behind their late start or late finish dates.
2. Substantial delays of resource availability are indicated.
3. The time durations of future activities appear to have been materially underestimated.

4. Logic changes in work yet to be performed have become necessary.

Where some form of corrective action is indicated, it must be done promptly if it is to achieve its objective.

Corrective action to bring a construction project back on schedule is of the same type previously discussed in Chapter 7 with reference to project time reduction. The critical path or paths can be shortened by compressing individual activities or through localized reworking of the network logic. As before, this often involves the testing of possible alternative courses of action. If there has been no recent updating of network calculations, the analysis of the weekly progress reports gives management only a general idea of the time status of the job. Consequently, the efficacy of a given corrective measure can be known only approximately unless it is simulated by incorporating it into a network update.

The entire process of project time management can be enhanced considerably by scheduling periodic job progress meetings. Held weekly, biweekly, or monthly, such site conferences are attended by project management, field supervisors, major subcontractors, material suppliers, owner representatives, and other parties as appropriate. Although such meetings can and do serve as a forum for a variety of job topics, the time status of construction operations is always a major consideration at such sessions. Face-to-face discussions can be very productive in eliciting ideas and obtaining cooperation from those who play major roles in keeping the work on schedule.

At such a job progress meeting on the highway bridge, it has been decided that some form of corrective action must be devised that will remedy the job delay occasioned by the difficulty with activity 140. Discussion reveals that considerable time recovery could be possible by making a network logic change involving construction equipment. This change would involve removing activity 160, "Demobilize pile driving rig," as a predecessor to activity 180, "Forms & rebar, abutment #1." Previous discussion in Chapter 5 required this sequence so that a crane would be available for the placing of forms and rebar for abutment #1. The decision is now to eliminate this equipment dependency by providing another somewhat smaller crane for the forming and placing rebar of abutment #1. This would remove activity 160 as a necessary prerequisite for activity 180. For now, activity 160 will be made a prerequisite for activity 240, "Forms & rebar abutment #2." This logic change can be affected with little additional expense. The effect of this job logic change will be determined in Section 9.16.

9.15 NETWORK UPDATING

As construction proceeds, diversions from the established plan and schedule inevitably occur. Unforeseen job circumstances result in changes of activity durations, activity delays, and changes in project logic. Remedial actions such as resource leveling and time expediting can produce similar effects. As such deviations occur and accumulate, the true job status diverges further and further from that indicated by the programmed plan and schedule. At intervals, therefore, it becomes necessary to incorporate the changes and deviations into the working operational program if it is to continue to provide realistic management guidance. This is accomplished by a procedure called "network updating."

The basic objective of an update is to reschedule the work yet to be done using the current project status as a starting point for the redetermination. Updating reveals the current time posture of the job, indicates whether expediting actions are in order, and provides guidance as how best to keep the job on schedule. An update is also very valuable in testing the effectiveness of proposed time recovery measures.

Updating involves making necessary network corrections and recomputing activity and float times. It is concerned entirely with determining the effect of schedule deviations and plan changes on the portion of the project yet to be constructed. Included here are not only the unexpected departures from the program, but also those corrective actions initiated to remedy specific time progress and resource availability problems.

From the viewpoint of time-management effectiveness only, an update as of each weekly progress report could be advantageous. This would keep job management continuously up-to-date on the time status of the work and would ensure prompt and informed remedial action when it is needed. However, this can involve considerable effort and expense. No definite rule exists about how often a network update should be made. The need for recomputation depends more on the seriousness of the plan and schedule deviations than on their number. There is a point of diminishing returns in retaining an outdated plan and schedule. Attempting to make the project fit an obsolete program is an exercise in futility and literally does more harm than good. Time control, to be optimally effective, must be based on a correct and current job model. What is important is not how often the network is recalculated, but how well the plan and schedule continue to fit the actual conduct of the work.

Updating computations are normally performed by computer although manual computations are discussed in Section 9.16 to provide a complete understanding of the process involved. Before an update can be made, the network must be corrected to reflect the latest information concerning the logic and durations of all work that is yet to be performed. This is an important point. The activities that have been completed are now history, and they are immutable. To determine the time condition of the remaining activities, the new computations must start from the current project status as of the designated data date.

To make a network update, the following information as of the cut-off date is required concerning the work that has yet to be completed:

1. New activities that must be added to the network.
2. Existing activities that are to be deleted.
3. Changes in job logic.
4. Changes in original material delivery or other resource availability dates.
5. Estimated times to complete all activities presently underway, but not yet completed.
6. Changes in estimated activity durations.
7. Changes in the scope of the work.

With regard to information needed for a project update, subjective input from the field supervisors and job expeditor can be especially valuable. For instance, the project supervisor can furnish revised information about future progress based on recent job experience. This information may relate to expected activity durations or to future logic changes. The expeditor may be able to input information regarding labor disputes, business disruptions, and other pertinent factors pertaining to supplier plants and transportation media. It may be possible to combine material shipments for one job with those of another to expedite deliveries, conditions that were unforeseeable when the original plan was devised. All this information needs to be reflected in the updated schedule.

Figure 9.7 summarizes all the information needed for updating the highway bridge as of the afternoon of July 21. The first four items have been previously discussed in conjunction with the weekly progress report in Figure 9.5. Item 5 is a revision upward in the estimated duration of a future activity. Item 6 was discussed in Section 9.14.

9.16 MANUAL UPDATING CALCULATIONS

To perform a manual update on the highway bridge as of July 21 (working day 27), the information contained in Figure 9.7 must first be incorporated into the project network. Figure 9.8 is the corrected precedence diagram used for the recomputation of activity times and floats. In Figure 9.8, those activities that have been completed as of day 27 are those to the left of the diagram which are drawn in shaded lines. All of the completed work items show only the activity description and number. Activities in progress, but not yet completed, are indicated by the dashed line labeled with the notation "July 21, Day 27." For these activities, the time durations shown have been changed to the number of working days estimated to reach completion. The dashed line shows the present stage of advancement of the work as of day 27 and is often referred to as the "time contour."

A number of diagrammatic conventions other than that shown in Figure 9.8 can be used with regard to making manual updating calculations. For example, the

Item	Network Update Information as of July 21 (working day 27)
1	Add activity 65, Fabricate & deliver deck rebar.
2	Anticipated delivery date of deck rebar is August 9 (working day 40). Activity 65 requires 40 - 27 = 13 working days to complete.
3	The estimated time to complete activity 140 is 2 working days
4	Anticipated delivery date for the girders is August 4 (working day 37). Activity 260 requires 37 - 27 = 10 working days to complete.
5	The estimated time to accomplish activity 330, Deck forms & rebar, has been revised upward from 4 to 6 working days.
6	A 25 ton crane was brought in and used to place the forms and rebar in activity 180. This removed the dependency between activity 160 and 180. A dependency was established between activity 160 and 240.

Figure 9.7 *Highway bridge, network update information.*

durations of all completed activities can be set equal to zero and the durations of activities in process set equal to the days to complete. The start activity is replaced with an "Elapsed time" activity whose duration is set equal to the data date (27 in this case). Computations are started at the beginning of the network by setting the early start of the elapsed time box equal to zero. Another scheme is to drop all completed activities from the network entirely. Activities left without predecessors are connected directly to a start activity whose early finish is set equal to the data date. The durations of partially completed activities are again set equal to their remaining durations. Actually, it is easy enough to start the forward pass directly at the time contour. This has been done in Figure 9.8 simply by entering an EF of 27 for the completed activities at the time contour and an ES of 27 for the activities in process.

The recomputation in Figure 9.8 proceeds as a normal forward pass followed by a backward pass. The forward pass discloses that, as of day 27, the project completion time will be 71 working days. This would indicate that the job is now only one day behind its original schedule. Thus, the equipment logic change discussed in Section 9.14 designed to ease the job delay caused by the problems encountered by the pile driving on abutment #2 (activity 140), has proven to be effective. The job delay has now been reduced from five working days to one day. The decision is now to accept the one-day delay and make the backward pass in Figure 9.8 starting with a project duration of 71. The critical activities are indicated by heavy-lined boxes and the critical path by heavy dependency lines. Comparing Figure 9.8 with Figure 5.1 reveals that the initial portion of the critical path has moved from its original position. Such changes in the identification of critical activities are common occurrences as the work progresses in the field. If a project duration of 70 is to be maintained activity 410, "Contingency," would be reduced to five days and the computed activity times would remain as before. What turn-around values are used on update calculations vary with personal preference. The matter is unimportant because there is no substantive difference in the management data generated.

Updated networks, such as that in Figure 9.8, can also be used to record the actual start and finish dates of each activity. Noting these dates above each activity box provides an historical record of actual job progress. This action is not usually required where job bar charts are maintained or where computer procedures are being used because it only duplicates information available elsewhere. A file of the successive update networks should be kept as a record of how the job progressed during the contract period.

9.17 COMPUTER APPLICATION

Schedule updating is one of the most useful and important computer applications in the entire project management system. The computer can provide timely management information in an easily understood and immediately usable form. Several computer programs are available that generate a wide variety of time-status reports. The field progress information is entered on computer-generated forms whose preparation requires minimum effort. Computer software is highly flexible in that the computer can deliver just about any type of report the project manager wants. Reports can be selectively prepared for any level of management or supervision. From summary statements generated for top management to highly detailed repots for field supervisors, the computer can produce time-status information for any desired level of job management.

At the field supervisory level, computers can be programmed to produce a wide variety of useful data expressed in terms of calendar days and dates. Typical information generated might be the following:

1. Estimated duration and actual duration of each completed activity.
2. Scheduled start dates and actual start dates of all activities completed or in progress.
3. Scheduled finish date and the actual finish date of each completed activity.
4. Time status of each activity in progress showing the anticipated finish date and the number of days ahead or behind schedule.
5. A revised project completion date with an indication of the projected time overrun or underrun.
6. Originally scheduled and revised start and finish dates for each activity not yet begun.
7. Identification of critical activities.
8. Float values of all activities.

This information can be generated using any desired date as the starting point for the backward pass.

9.18   SCHEDULE INFORMATION ON THE JOB

Although the project manager is responsible for the overall application and direction of the project time-management system, the field supervisors also play key roles in keeping the project on schedule. It is the field supervisors who put into action the plans and schedules devised by project management. Consequently, if project time management is to work, there must be some established means of communication between office and field. The project manager must keep his field supervisors currently and accurately informed concerning the schedule of operations and the time status of the work.

The information transfer cannot be accomplished by merely relaying piles of computer printout to the supervisors who have neither the time nor the inclination to search out the information relevant to their individual responsibilities. The project manager must provide concise, short-range schedule information that will meet the specific needs of each recipient. Limited to pertinent subject matter and in a level of detail appropriate to the user of the information, the data provided should quickly and clearly communicate schedules, current time status, scheduling leeway, and trouble spots.

9.19 PROJECT PROGRESS CURVES

This chapter has discussed the time or progress monitoring of construction projects in the context of comparing the actual progress of job activities with that planned. This procedure provides detailed information concerning the current time status of individual job segments. Overall job progress is also used as a time-monitoring device, either as a stand-alone system or in conjunction with the detailed procedures already discussed. Total project progress as of a given date can be expressed in terms of different cumulative measures such as total labor cost, total money expended, work quantities put into place, total man-hours used, or possibly others. Actual numbers of units or percentages of the total can be used. Progress curves are prepared by plotting cumulative job progress, expressed in terms of one of the measures just cited, to a horizontal time scale. Progress monitoring is accomplished by periodically comparing planned project progress with actual values.

A planned progress curve is obtained by calculating cumulative totals of the chosen progress measure at the end of each successive time unit. Depending on the nature and extent of the work, days, weeks, or months can be used in the preparation of progress curves. If the progress values are plotted at short time intervals, the resulting curves are apt to be irregular and jerky in appearance. Plotting values as of the end of each month normally results in reasonably smooth curves. A common and very effective procedure in the plotting of planned progress curves is to produce two such curves. One is determined on the basis of all project activities starting as early as possible. The other is based on all project activities starting at their late start dates. When these two curves are plotted, they form a closed envelope. Figure 9.9 shows such an envelope plotted in generalized form. These curves are commonly referred to as "S-curves" because of their typical appearance.

After two extreme condition curves are drawn, an average curve between them is sketched in. This average curve is shown as a dashed line in Figure 9.9 and is used for purposes of general progress monitoring. During the construction period, actual progress is plotted periodically, these points forming the actual progress line. The relative position of this line to the planned average progress curve is used to evaluate the time status of the project as a whole. Where actual progress plots above the average progress line, the time status of the job is considered to be satisfactory. When it lies below, time progress is considered to be generally unsatisfactory.

A study of the general geometry of a project envelope can provide interesting information concerning the work at hand. Figure 9.9 illustrates the general form of this envelope for a typical construction project. A rule of thumb often used is that 50 percent of the job is accomplished during the middle one-third of the construction period with the other 50 percent being about equally divided between the initial and final thirds. The location of the ES and EF curves with respect to one another depends on the relative amounts of activity float present. If the floats tend to be small, the two curves will be close together, and the shape of the envelope will be long and thin. If relatively large amounts of float are present, the two curves become more widely separated. Consequently, the shape of the envelope gives a quick visual indication of the degree of time control required to keep the project on schedule. A narrow envelope will require a considerably more rigorous time-control program than will a project where the envelope indicates the

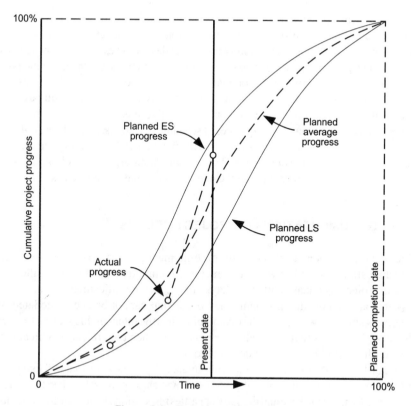

Figure 9.9 *Project progress curves.*

presence of substantial float. A narrow envelope will also present more difficulties in resource allocation because of the limited float.

Project progress curves, in themselves, are of limited effectiveness insofar as presenting the time status of a construction project is concerned. Although they do indicate the time status of the project as a whole, they are of no assistance in solving time-slippage problems. However, for purposes of the owner, architect-engineer, and top project management, such curves can be very useful in affording a quick grasp of the overall time condition of the work. For this reason, progress curves are frequently superimposed on the project bar chart. In this way, all the progress information is brought together into a single graphical display.

9.20 EXTENSIONS OF TIME

Construction contracts typically provide that the contractor shall complete all work on the project within a designated period of time. This is a very important contract provision and the contractor will be in breach of contract if it does not meet the prescribed time requirement. There are many sources of delay that invariably occur during the contract period. Whether any of these entitle the contractor to an extension of time depends on the specific provisions of the contract.

Requests for extensions of time must be initiated by the contractor as provided for by contract, and the burden of proof in justifying such a contract change lies with the contractor. Documenting responsibility for project delays and establishing the lengths of time involved can be an arduous task, but is one of extreme importance for the project manager. Not only extensions of contract time are involved here. Establishing who was responsible for a work disruption or delay will usually determine who will pay for the attendant costs. The project network is an effective mechanism for analyzing and understanding the ultimate consequences of schedule deviations. Not only is the diagram a valuable analytical device, it can also serve the function of documenting pertinent facts pertaining to work changes and delays. The most important aspect in resolving associated disputes is careful documentation of the pertinent facts and circumstances.

9.21 DOCUMENTATION OF PROJECT DELAYS

The previous sections of this chapter on time management have discussed project schedule slippages and how the contractor accommodates them within the established project calendar. Unfortunately, such schedule adjustment is not always possible and major time problems may occur that cannot be accommodated by a relatively simple reworking of the overall plan. It is the intent of this section to discuss the kinds of actions that the contractor must take when the project is subjected to a serious disruption involving major time loss.

When the contractor is faced with a substantial delay in the work that will likely cause project completion to run over contract time, immediate steps must be taken to document the reason for its occurrence and the party or parties responsible. If the contractor is at fault, such a breach of contract can lead to the assessment of costs against the contractor, these being either in the form of liquidated damages provided for by contract or damage imposed by arbitration or the courts. In cases where the delay is caused by the owner, subcontractor, material dealer, or other party, the contractor may have to sue the offending party for damages and prove its case in court. In either event, the contractor must be able to document the time aspects of the delay and demonstrate the roles played by the responsible parties. In this regard, a suitably maintained project network is an excellent medium for analyzing the short- and long-term effects of a serious time problem and for demonstrating responsibility for schedule disruptions and loss of time.

When planning and scheduling are so disrupted, the principal objectives are to reschedule the project with minimum impact on schedule and cost, and to establish responsibility for the delay. An excellent way to do this is through the use of a suitably revised network showing the necessary changes and the attendant activity recomputations. The following example illustrates such a procedure.

Assume that on the afternoon of July 14, project day 22, the status of the bridge project is just as shown in Figure 9.4. However, the trouble with the pile driving on abutment #2, activity 140, has become much more serious than described in Section 9.11 and the pile driving is making little progress. During the afternoon of July 14, the owner's engineer directed the contractor to stop work on the pilings for abutment #2 until the problem could be studied. All other work can proceed as planned. As shown in Figure 9.10, the contractor now inserts a milestone entitled "Stop order on piledriving, abutment #2" and adds two new activities. Activity 155 shows that the

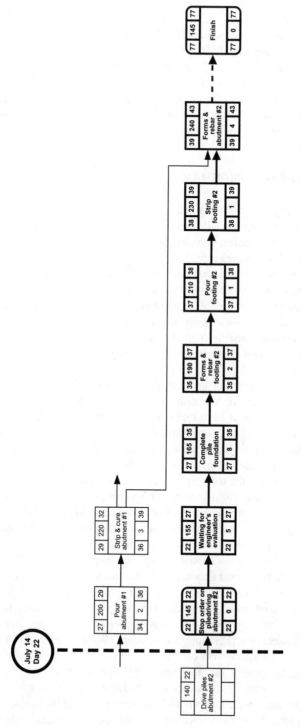

Figure 9.10 Revised network, problem with activity 140.

engineer required five working days to devise a suitable solution. Upon receipt of the engineer's directive in this regard, the contractor removed the troublesome material and replaced it with compacted backfill. The required pilings were then driven, the entire process requiring a total of eight working days as shown by activity 165. This information is shown in Figure 9.10 and the forward pass computations are completed to project completion on this basis. Figure 9.10 shows that the project now has an anticipated finish time of 77 working days, six days later than indicated by Figure 9.8. It is to be noted that Figure 9.10 clearly shows the cause of the delay and its effect upon the time for project completion.

9.22 THE AS-BUILT DIAGRAM

Very useful with regard to the settlement of contract disputes involving time is a project network variously referred to as an "as-built" diagram or "factual" network. This diagram is drawn as the work progresses and provides a comprehensive record of the work performed to date. This network is made to show each work change and delay as it occurs, together with the cause, names, and pertinent dates. The network also records the actual start and finish dates of every activity. To accomplish their purpose, diagrams of this type must be prepared as the work evolves, not long after it has been accomplished. When supplemented with written narrative, an as-built diagram provides an accurate and detailed history of the construction period. These networks are frequently drawn to a time scale, thus enhancing their effectiveness in providing a clear and quantitative picture of the total consequences of a given project change or delay.

The usual periodic network updatings do not normally serve to show the time effects of any one network variation. Even though updatings do start from the equivalent of the current as-built diagram, they normally determine the collective effect of several accumulated network deviations. Consequently, it is impossible to assess the effect of any single network change from the data generated by a routine diagram recalculation. However, the as-built diagram can be updated at any stage of the project to determine the total time effects of any given network variation. Such a recalculation will clearly illustrate the effect that any one delay or change would have or did have on the project completion date. Thus it is possible to ascertain the basic causes of project completion delays and to attach responsibility for such delays in accordance with contract provisions.

With regard to extensions of contract time, it is important to note that current judicial opinion often holds that a contractor is not entitled to a time extension when a change or delay only consumes float time. This concept is reinforced by the inclusion of such provisions in some contractual documents. This brings up the interesting question of just how real float time is in conventional construction networks. Previous discussions in this text relative to project time shortening and resource allocation indicate that much of the float indicated by a normal network calculation is illusory and really does not exist. As the work proceeds, the project manager and field supervisors use such spare time to accommodate a constant series of factors not included or recognized by the usual planning and scheduling procedures. Float is not subject to the whim of the project manager, but is carefully distributed over the project to smooth resource requirements and make the job run as efficiently as possible.

10

THE PROJECT
COST SYSTEM

10.1 OBJECTIVES OF COST SYSTEM

The project cost system is designed to accomplish two important objectives.* One is to develop labor and equipment production information in a form suitable for estimating the cost of future work. The ability to estimate construction costs accurately is a key element in the success of any contracting firm. The other objective of a project cost system is to keep the construction costs of the project within the established control budget. Regardless of the type of contract with the owner, it is important that the contractor exercise the maximum control possible over its field costs during the construction period. A functioning and reliable cost system plays a vital role in the proper management of a construction project.

How the costs of a construction project are controlled is a variable with its size and character. A large, complex job requires a detailed reporting and information system to serve project management needs. Simpler and less elaborate cost systems are sufficient for smaller and simpler projects. In any event, the only justification for the expense of a project cost system is the value of the management data that it provides. If the information produced is not used or if it is not supplied in a usable form or timely fashion, then the cost system has no real value and its cost cannot be justified. Properly designed and implemented, a project cost system is an investment rather than an expense.

This chapter discusses the methods involved in project cost control and the estimating feedback process. Although the details of how these actions are actually accomplished vary substantially from one construction firm to another, the ensuing

---

* Portions of this chapter have been adapted from Richard H. Clough, *Construction Contracting*, 5th ed., New York: Wiley, 1986. Chapter 12.

treatment can be regarded as being reasonably typical of present practice within the construction industry.

10.2 PROJECT COST CONTROL

Project cost control actually begins with the preparation of the original cost estimate and the subsequent construction budget. Keeping within the cost budget and knowing when and where job costs are deviating are two factors that constitute the key to profitable operation. As the work proceeds in the field, cost accounting methods are applied to determine the actual costs of production. The costs as they actually occur are continuously compared with the budget. In addition to monitoring current expenses, periodic reports are prepared that forecast final project costs and compare these predicted costs with the established budget. Field costs are obtained in substantial detail because this is the way jobs are originally estimated and also because excessive costs in the field can be corrected only if the exact cause can be isolated. To learn that construction costs are going over the budget is not helpful if it is impossible to identify where the trouble is occurring.

Cost reports are prepared at regular time intervals. These reports are designed to serve as management-by-exception devices, making it possible for the contractor to determine the cost status of the project and to pinpoint those work classifications where expenses are excessive. In this way, management attention is quickly focused on those job areas that need it. Timely information is required if effective action against cost overruns is to be taken. The detection of excessive costs only after the work is finished leaves the contractor with no possibility of taking corrective action.

10.3 DATA FOR ESTIMATING

As discussed in Chapter 3, when the cost of a project is being estimated, many elements of cost must be evaluated. Labor and equipment expenses, in particular, are priced in the light of past experience. In essence, historical production records are the only reliable source of information available for estimating these two job expenses. The company cost system provides a reliable and systematic way of accumulating labor and equipment productivity and costs for use in estimating future jobs.

With regard to feedback information for cost-estimating purposes, there is some variation in practice about the form of production data that are recovered. It can be argued that production rates, previously discussed in Section 2.10, are fundamental to the estimating of labor and equipment costs. However, as seen in Chapter 3, costs per unit of production or "unit costs" are widely used for estimating labor and equipment because of the convenience of their application. Such unit costs are, of course, determined from production rates and hourly costs of labor and equipment. Such unit costs can be kept up-to-date by being adjusted for changes in hourly rates and production efficiencies.

Information generated for company estimating purposes can, therefore, be in terms of labor and equipment production rates, unit costs, or both. The difference in usage seems to be at least partially a matter of the extent of a contractor's operations. Smaller contractors frequently work exclusively in terms of unit costs while larger companies must, almost of necessity, base at least portions of their estimating on production rates. Both procedures were used in Chapter 3 for purposes of estimating various segments of the highway bridge. It is obvious that the feedback system to estimating must be designed to produce information in whatever form or forms are compatible with company needs and procedures.

10.4 PROJECT COST CODE

Each individual account of a contractor's financial accounting system is assigned its own code designation as a means of classification and identification. Here, only the cost codes for job expense accounts are involved. Many construction firms use their own customized project cost codes that they have developed and tailored to suit their individual preferences and needs. Many forms of alphabetical, numerical, and mixed cost codes are in use.

A number of suggested standard coding systems have been developed for different types of construction work by various trade and professional groups. The cost code used here is based on the Broadscope Section Titles of MASTERFORMAT that was jointly produced by The Construction Specifications Institute (CSI) and Construction Specifications Canada (CSC). The Broadscope Section Titles and Numbers are used along with a discussion of the project cost system and are reproduced in Appendix D.

The project cost code for the highway bridge is derived from the section titles and numbers in Appendix D. Each work type involved on this job is assigned its unique numerical designation. However, the code number taken from Appendix D constitutes only one part of the total identifying account number for each work type because a project code typically includes other kinds of information. To illustrate this point, each item of expense that is incurred on the highway bridge will carry a 14-digit cost code identification. An example might be:

$$9108\ 05\ 03157.20.3$$

This cost code number is the labor cost of placing and stripping the abutment forms on the highway bridge. The cost code number contains these data groups:

1. *Project number.* The first four digits identify the particular project on which the cost is incurred and to which the cost is to be charged. Different construction companies have their own ways of doing this. Here, the Example Project is designated as Project No. 9108, meaning the eighth project started in 1991. If desired, other information can be included such as the form of contract (unit price) and kind of work (heavy construction).

2. *Area code.* The area or location within the project is indicated by the next two digits in the cost code number. This concept applies only to large projects where there are usually distinctive features that naturally divide the work into

separate geographical areas or major physical parts. The Example Project consists of several major segments, including the dam, highway bridge, pipeline relocation, and others. In this case, the highway bridge is assigned an area code designation of 05. One of the values of the area code is that it makes it possible to associate field costs with the specific supervisors and managers responsible for them.

3. *Work type code.* This seven-digit number comes after the area code and is acquired from Appendix D. The section titles and numbers in Appendix D are the master accounts with specific work-type accounts being derived from the master list. To illustrate, if the type of work involved is placing and stripping abutment concrete forms, reference to Appendix D shows that forms are assigned the number 03100. Code numbers for specific forming systems and materials are obtained by changing the last digit in the master account number. For purposes of this chapter, wooden forms are designated by 03157. Project account numbers to indicate different form applications on the highway bridge are obtained by placing a decimal point after the number 03157 and adding digits to the right of the decimal point. Abutment forms are correspondingly identified by 03157.20 (footing forms are 03157.10 and deck forms are 03157.30).

4. *Distribution code.* To serve a variety of accounting purposes in addition to that of project cost accounting, it is necessary to indicate the category of expense involved. This is done by means of a standard company distribution code. For example, "1" is total, "2" is material, "3" is labor, "4" is equipment, and "5" is subcontract. Thus, the final numeral 3 in this example cost code indicates that it is a labor charge. With regard to the distribution code, "total" merely indicates an all-inclusive cost that may include any combination of labor, material, equipment, or subcontract cost.

10.5 USAGE OF PROJECT COST CODE

Labor, materials, supplies, equipment charges, subcontract payments, overhead costs, and other items of expense are charged to the project where they are incurred. To ensure that each such expenditure is properly charged to the correct cost account, every expenditure is coded in accordance with the project cost code system. The use of "general" or "miscellaneous" cost accounts is poor practice and is to be avoided. The job expense accounts provide the basic information for the periodic project cost reports that will be discussed subsequently.

If the project cost code is to serve its basic purpose, it is necessary that it be understood and used consistently by all company personnel. It is mandatory that the project be broken down into the same established work types and that the project cost code be used consistently for the purposes of project estimating, general financial accounting, and field cost accounting. When the estimator prices a new project, the work is subdivided into the standard classifications. Reference to Figure 3.6 and Appendix D shows that the estimator identifies each work quantity by its work type code when he enters the result of the quantity take off on the summary sheets. The project number and area code are assigned at a later time. From initial estimate to project completion, the same cost code numbers and elementary work types are used. Although the project number and area code are unique to a given job, the work-type code and distribution code are used consistently on all projects. An important point here is that each cost account must consistently contain the same elements of cost. For example, labor costs must always either include or not include applicable indirect costs such as fringe benefits and payroll taxes and insurance.

In the usage of day-to-day cost forms and in the preparation of periodic project cost reports, the project number and area code are not usually included as a part of the individual cost-account numbers. Rather, they appear only in the heading of the form or report. This is also true when cost records of completed projects are filed. Permanent files of cost records from past projects are a very valuable estimating resource. The fact that historical job costs can be associated with specific projects makes them even more beneficial. This enables the user of the cost files to associate production rates and costs with specific project circumstances and conditions.

10.6   PROJECT COST ACCOUNTING

Project cost accounting is the key ingredient in the project cost system. It provides the basic data required for both cost control and estimating. Cost accounting differs substantially from financial accounting, however. Cost accounting relates solely to determining the detailed makeup of productivity and costs associated with the production of a construction product in the field, including the necessary overhead expense. Project cost accounting is not separate from the contractor's general system of accounts. Rather, it is in the nature of an elaboration of the basic project expense accounts. Cost accounting involves the continuous determination of productivity and cost data, the analysis of this information, and the presentation of the results in summary form.

It is thus seen that project cost accounting differs from the usual accounting routines in that the information gathered, recorded, and analyzed is not entirely in terms of dollars and cents. Construction cost accounting is necessarily concerned not only with costs, but also with man-hours, equipment-hours, and the amounts of work accomplished. The systematic and regular checking of costs is a necessary part of obtaining reliable, time-average production information. A system that evaluates field performance only spasmodically in the form of occasional spot checks does not provide trustworthy feedback information, either for cost control or for estimating purposes.

Project cost accounting must strike a workable balance between too little and too much detail. A too-general system will not produce the detailed costs necessary for meaningful management control. Excessive detail will result in the objectives of the cost system being obliterated by masses of data and paperwork as well as needlessly increasing the time lag in making the information available. The detail used in this book with regard to the estimating and cost control of the highway bridge is reasonably typical of actual practice in the industry.

A project cost-accounting system supplements field supervision; it does not replace it. In the final analysis, the best cost-control system that a contractor can have is skilled, experienced, and energetic field supervision. It is important that field supervisors realize that project cost accounting is meant to assist them by the early detection of troublesome areas. Trade and site supervisors are key members of the cost-control team and without their support and cooperation, the job cost system cannot and will not perform satisfactorily.

10.7   LABOR AND EQUIPMENT COSTS

It is a basic accounting principle for construction contractors that project income and expense are recorded by individual job. The contractor's financial accounting system includes project expense accounts that are used to record every item of expense charged to any given project. Job costs associated with materials, subcontracts, and nonlabor items of project overhead are of a reasonably fixed nature, and cost control of these kinds of expenses is effected mainly by disbursement controls applied to purchase orders and subcontracts (Section 11.15 provides more on this). Barring oversight or mistake during the estimating process, these costs are determined with reasonable exactness when the job is priced, and such costs seldom tend to vary appreciably from their budgeted amounts. For this reason, the cost information available from the monthly cost forecast reports (see Section 10.21) pertaining to materials, subcontracts, and nonlabor field overhead items is normally suitable and timely enough for ordinary cost-control purposes.

Labor and equipment costs are an entirely different matter, however. These two categories of job expense are characterized by considerable uncertainty and can fluctuate substantially during the construction period. These are the only categories of job expense that the contractor can control to any extent, and they merit and need constant management attention. However, the usual monthly cost forecast reports do not normally suffice for labor and equipment cost control. This is because the information concerning these two categories of field expense is not reported in adequate detail or at sufficiently short time intervals. Detailed cost-accounting methods must be used in conjunction with labor and equipment expense if effective management control over them is to be realized. Accordingly, the discussion of project cost accounting in this chapter is essentially limited to these two categories of job expense. Described will be the process of determining at regular intervals how much work is being accomplished in relation to the amounts of labor and equipment being used.

10.8   COST ACCOUNTING REPORTS

Summary labor and equipment cost reports must be compiled often enough so that excessive project costs can be detected while there is still time to do something about them. Cost report intervals are very much a function of project size, nature of the work, and the type of construction contract involved. Obviously, there must be some kind of balance struck between the cost of generating the reports and the value of the management information received. Daily cost reports are frequently prepared on complex projects involving multiple shifts. It is not the usual job, however, that can profit from such frequent cost reporting. Some very large projects that involve relatively uncomplicated work classifications find monthly or even longer intervals to be satisfactory. For most construction projects, however, cost reports are needed more often than this.

It is generally agreed that weekly labor and equipment cost summaries are about optimum for most construction operations, and this is the basis for discussion here. The cutoff time can be any desired day of the week, although contractors will often match their cost system to their usual payroll periods and also to their time-monitoring system. This matter has previously been discussed in Section 9.10. The cutoff times used here for weekly labor and equipment cost reports will be Wednesday afternoons. These are the times it is assumed payroll periods end and weekly work quantity measurements are made.

10.9   LABOR TIME REPORTING

The source documents for labor costs and cost code allocations are labor time cards, the same cards used for payroll purposes. They report the hours of labor time for every tradesman and the project cost codes to which the labor applied. Figure 10.1 shows a typical daily time card and Figure 10.2, a weekly time card. Which of these two is used depends on company preference and internal routines. Many construction companies' standard operating procedures require the use of daily time cards. For example, some companies with computer-based payroll systems will, each day, run the previous working day's time cards through a preliminary processing routine. Other companies prepare labor cost reports on a daily basis.

When daily time cards are used, they are filled in and forwarded each day to the company's payroll office. The head of the card provides for entry of the project name and number, date, weather conditions, and the name of the person preparing the card. The body of the time card provides for the employee number, name, and craft for each person covered. In Figures 10.1 and 10.2, the designation "F" indicates foreman, "C" indicates carpenter, "L" means laborer, and "O" denotes equipment operator. Hours are reported as straight time (ST) or overtime (OT) as the case may be. For each person listed, several slots are provided for the distribution of hours to specific cost codes. Absolute precision in allocating labor time is not possible. Nevertheless, the need for care and reasonable accuracy in ascribing each person's time to the appropriate cost codes cannot be overemphasized.

Contractors vary in how they charge their craft supervisors' time. Some prefer to charge it directly to the work classifications on which the supervisor spent time. Others charge all foreman time to a supervisory account in project overhead. Probably the most common procedure is to allocate the time of craft supervisors who work with their tools to the work-type account. Otherwise, the time is charged to job overhead. If weekly time cards are used, a separate card for each worker is prepared to record the hours and cost codes for that week. Although the diagrammatic arrangement of the weekly time card is different from that of the daily time card, it contains the same payroll and cost accounting information concerning the individual tradesman.

DAILY LABOR TIME CARD

Project Highway Bridge Project No. 9108.05 Weather Cloudy-windy
Date July 20   Prepared by R.H.C.

Employee Number	Name	Craft	Time Classification	Hourly Rate	Cost Code 0315710.3	Cost Code 0315720.3	Total Hours	Gross Amount
132	Winnowich, N.	F	ST	$17.20		8	8	$137.60
			OT					
221	Clouten, S.	C	ST	$16.20		8	8	$129.60
			OT					
248	Schluder, L.	C	ST	$16.20		8	8	$129.60
			OT					
319	Mills, R.	O	ST	$15.95		2	2	$31.90
			OT					
143	Gibson, E.	L	ST	$11.60		8	8	$92.80
			OT					
417	Sears, K.	L	ST	$11.60		8	8	$92.80
			OT					
176	Sibanda, M.	L	ST	$11.60	2	6	8	$92.80
			OT					
	Total Hours				2	48	50	
	Total Cost				$23.20	683.90		$707.10

Figure 10.1 Highway bridge, daily labor time card.

WEEKLY LABOR TIME CARD

Craft _____ C _____

Name Clouten, S.
Employee No. 221
Week Ending July 21

Project Highway Bridge
Project No. 9108.05
Prepared by R.H.C.

Cost Code	Time Classification	Hourly Rate	Thursday	Friday	Monday	Tuesday	Wednesday	Total Hours	Total Cost
03157.10.3	ST	$16.20	8					8	$129.60
	OT								
03157.20.3	ST	$16.20		8	8	8		24	$388.80
	OT								
03311.10.3	ST	$16.20					8	8	$129.60
	OT								
	ST								
	OT								
	ST								
	OT								
Total Hours	ST		8	8	8	8	8	40	
	OT								
Gross Amount			$129.60	$129.60	$129.60	$129.60	$129.60		$648.00
Weather			Clear-hot	Clear-hot	Cloudy-gusty	Cloudy-windy	Cloudy-windy		

Figure 10.2 *Highway bridge, weekly labor time card.*

10.10   TIME CARD PREPARATION

The distribution of each worker's time among cost codes is normally the responsibility of the field supervisor (foreman), who is in the best position to know how each worker's time was actually spent. It is usual that this information is first recorded in the foreman's pocket time book with the work performed by the individuals in his crew often being described in word form rather than by code numbers. Ordinarily, the foreman enters the hourly rate for each worker because it is not unusual for a person during a week, or even in a single day, to be assigned to work requiring different hourly rates of pay. However, extensions of hours and wage rates into totals are not usually made by him. The hourly rates shown on the time cards in Figures 10.1 and 10.2 are base wage rates only and do not include any indirect labor costs, such as payroll taxes, insurance, or fringe benefits. Labor unit costs derived from such time card information commensurately do not include any indirect labor costs either, a point dealt with in Section 3.15. In this text, all labor unit costs are derived from basic wage rates only.

The importance of accurate and honest time reporting cannot be overemphasized. On the basis of the allocation of labor (or equipment) time to the various account numbers, cost and production data are generated. If this information is inaccurate or distorted, it can be seriously misleading when used for estimating and cost-control purposes. Loss items must be identified as such without any attempt at coverup by charging time to other cost accounts whose performance has been good.

The formal time cards are frequently filled in by someone other than the foreman, such as the project timekeeper, cost engineer, or project manager. The timekeeper, for example, can collect the foremen's time books at some convenient time each day and fill out the time cards, adding the necessary information for payroll and cost accounting and making the extensions.

Even if weekly time cards are used, it is preferable that the labor distribution be made each day. Setting down the time and its allocation to different cost codes on a daily basis eliminates the undesirable practice of the foreman letting the matter go until the end of the week and then trying to enter the information from memory. Completing the labor record at the end of every working day will materially improve the accuracy of the distribution. It is for this and other reasons that many contractors favor the use of daily time cards.

10.11   MEASUREMENT OF WORK QUANTITIES

To determine productivity rates and unit costs, it is necessary to obtain not only the hours and costs expended, but also the amounts of work put into place. On some types of work, it may be feasible and convenient for the field supervisors to report work quantities accomplished as of the end of each day or each shift. It is more common practice, however, that project work measurements be made at longer intervals, such as at the end of each weekly payroll period, the basis for the discussion here. Although labor costs are now being discussed, the weekly measurement of work quantities includes all work items performed, whether accomplished by labor, equipment, or a combination of the two. Consequently, the same weekly work quantity determination serves for both labor and equipment cost accounting.

Work quantities can be obtained in a variety of ways depending upon the nature of the work involved and company management methods. Direct field measurement, estimation of percentages completed, computation from the contract drawings, use of the estimating sheets, and determination from network activity progress reports are all used. Direct measurement on the job is common. This procedure is simple and direct on projects that involve few cost code classifications. Many contracts of the heavy, highway, or utility category are of this nature. There are often instances where quantities can be approximated with reasonable accuracy by applying estimated percentages of completion to the total work quantities. Although not as accurate as direct measurement, this procedure can provide usable measures so long as more accurate determinations are made occasionally such as for monthly pay requests.

The field measurement of work quantities on projects that involve many cost code classifications can become a substantial chore. Most building and industrial projects entail substantial numbers of different cost cost classifications. One convenient procedure in such cases is to mark off and dimension the work advancement in colored pencil on a set of project drawings reserved for that purpose. The extent of work put into place can be indicated as of the end of each day or each week as desired. By using different colors and dating successive stages of progress, work quantities can be determined from the drawings or estimating sheets as of any date desired.

The field measurement of work quantities can be and often is done by the field supervisors. However, on large projects and especially those with many work codes, it may be desirable that the cost engineer or project manager carry out this function because of the time and effort required. Weekly reports of work done are submitted on standard forms such as that in Figure 10.3. Prepunched and preprinted cards with cost codes, work description, and unit of measure already entered can be used for reporting work quantities when cost reports are prepared by computer.

10.12 WORK QUANTITIES FROM NETWORK ACTIVITIES

Project network activities used for planning, scheduling, and time control can serve as convenient packages for determining work quantities accomplished in the field. The weekly progress report (see Figure 9.5) used for project time management submits information concerning completed activities and percentages of completion of activities in process. However, the activity completion information contained in the usual weekly progress report must be translated into cost code quantities (Figure 10.3). This can be readily accomplished by obtaining the work-type quantities associated with each activity from the estimating sheets or project drawings.

A note of caution here, however, is needed concerning the use of activities to determine work quantities accomplished. In the usual case, activity work quantities are taken from the project drawings. For most activities, quantities indicated by the drawings and those actually required in the field are essentially the same. However,

Cost Code	Work Type	Unit	Total Last Report	Total This Week	Total to Date
\multicolumn{6}{l}{WEEKLY WORK QUANTITY REPORT}					

WEEKLY WORK QUANTITY REPORT

Project Highway Bridge

Week Ending Wednesday, July 21 (working day 27)

Project No. 9108-05

Prepared by R.H.C.

Cost Code	Work Type	Unit	Total Last Report	Total This Week	Total to Date
02361.10	Piling, steel, driving	lf	1,456	560	2,016
03159.10	Footing forms, strip	sf	0	360	360
03157.20	Abutment forms, place	sf	0	1,810	1,810

Figure 10.3 *Highway bridge, weekly work quantity report.*

there may be a difference, at times, especially on unit-price contracts. When field quantities can differ to any extent from those indicated by the drawings, actual amounts of work items must be measured in the field.

When using activities as the basis for work quantity determination, the completed amounts of different work classifications can be readily and accurately established for those activities that have been completed. The situation may be considerably different, however, where activities in progress are involved. In the case of such an activity, its reported percentage of completion is based more on its remaining time to completion than it is on quantities of work physically in place. The usual assumption is that the rate at which work is accomplished is a linear variable with time. That is, if an activity is reported as being 30 percent complete, this is normally taken to indicate that 30 percent of the work units have been completed. For activities of limited extent and that involve only one cost code classification, this is usually a reasonable assumption.

However, linear work accomplishment with activity time is sometimes not even remotely true, especially with activities that entail more than one elementary work classification. For instance, on the highway bridge, there are activities that involve both forming and reinforcing steel placing. Such an activity is accomplished by the carpenters erecting the outside forms, the ironworkers tying the rebar, and the carpenters then putting up the inside forms and bulkheads. At the 30 percent point in time required for the activity, and this is what the usual progress completion report is based on, perhaps 45 percent of the carpenter work is now done, but none of the reinforcing steel has been placed. Written notes that routinely accompany the weekly progress repots will normally clarify such matters, however. Activities that involve more than one elementary work classification must be broken down into their component cost code classifications when the weekly work quantity report is being prepared.

10.13 COST RECORDS AND REPORTS

By what methods and in what forms the field cost and production data are recorded, analyzed, and reported depend on whether manual or computerized procedures are being used. If the cost system is maintained by hand, as can be the case with some small construction firms, some form of workbook is needed for the recording and analysis of the cost and quantity data. A separate record is kept of each cost account in which entry is made of all labor and equipment charges allocated to that account number and of all quantities of that work type accomplished in the field. These cost accounting worksheets are maintained separately from the usual project expense ledgers that are used for the recording of all items of job cost.

If cost procedures are computerized, as is true with most of today's construction contractors, labor and equipment cost reports are generated as an integral part of the overall payroll and accounting system. However, regardless of whether manual or machine methods are used, the basic data that go in and the summarized production and cost information that come out are essentially the same.

The detail used in cost reporting is adjusted according to the level of management for which it is intended. The cost information routinely provided to managers or supervisors must be tuned to the scope and nature of their job responsibilities. The site superintendent on the highway bridge would be provided with very detailed cost information concerning all aspects of his job. On the other hand, the project manager of the Example Project would receive only general information concerning major aspects of the highway bridge. Obviously, the project manager can request and receive additional details where the cost performance on some job item has been unsatisfactory.

10.14 WEEKLY LABOR REPORTS

Week labor reports can be prepared on either a man-hour or cost basis. That is, labor productivity can be monitored in terms of either man-hours per unit of work (production rates) or cost per unit of work (unit prices). Which of these is used is a function of project size, type of work involved, and project management procedures. Where man-hour control is used, a budget of man-hours per unit of work is prepared. In this regard, total man-hours are usually used with no attempt to subdivide labor time by trade specialty such as so many hours of carpenter time and so many of ironworker time. Total man-hour estimates are based on an "average" crew mix for each work type.

During field operations, actual man-hours and work quantities are obtained. This makes possible a direct comparison of actual to budgeted productivity. Such an approach, of course, reflects productivity, but not cost. It is simple to implement and avoids many problems associated with labor cost analysis. One such problem occurs on projects that require long periods of time to complete. Wage rates may be increased several times during the life of such a job. When the project cost is first being estimated, educated guesses are made about what the wage raises will likely be. What this means is that labor costs on long-term jobs are often estimated without exact knowledge of the wage rates that will actually apply during the construction process. As a result of this, actual labor wage rates during the work period may turn out to be different from those rates used in preparing the original control budget. Thus labor costs produced by the project cost system are not directly comparable to the budget. To make valid cost comparisons, it is necessary to either adjust these costs to a common wage-rate basis or work in terms of man-hours rather than labor cost.

In the foregoing case, man-hours can serve as a very effective basis for labor cost control. However, for most construction applications, including smaller projects such as the highway bridge, labor cost analysis is more widely applied than is man-hour analysis. For this reason, the labor cost reports discussed here are all based on costs.

10.15 WEEKLY LABOR COST REPORT

Once a week, labor costs obtained from the time cards are matched to the work quantities produced. The results of this analysis are summarized in a weekly labor cost report, two different forms of which are illustrated in Figures 10.4 and 10.5. These labor reports classify and summarize all labor costs incurred on the highway bridge up through the effective date of the report (July 21). The labor costs in these two reports are direct labor costs only and do not include indirect labor costs. These reports have the objective of providing job management with detailed information concerning the current status of labor costs and of indicating how these costs compare with those estimated. Both of the labor report forms are designed to identify immediately those work classifications having excessive labor costs and to give an indication of how serious those overruns are. Labor cost reports vary considerably in format and content from one construction company to another, although all such report forms are designed to convey much the same kind of management information.

Figure 10.4 is a comprehensive weekly cost report for the highway bridge that summarizes labor costs as budgeted, for the week being reported, and to date. Not all cost report forms include costs for the week being reported. These values can be of significance, however, in indicating downward or upward trends in labor costs. The labor report form in Figure 10.4 involves work quantities as well as labor expense and yields unit costs for each work type. Unit prices, obtained by dividing the total labor cost in each work category by the respective total quantity, enable direct comparisons to be made between the actual costs and the costs as budgeted. In Figure 10.4, the budgeted quantity, budgeted direct labor cost, and budgeted labor unit cost for each work type are taken from the project budget, Figure 3.10. The other quantities and labor costs are actual values, either for the week reported or to date.

When the total quantity of a given work item has been completed, its to-date and projected saving or loss figures are obtained merely by subtracting its actual total labor cost from its estimated total cost. When a work item has been only partially accomplished, the to-date saving or loss of that work item is obtained by multiplying the quantity in place to date by the underrun or overrun of the unit price. The projected saving or loss for each work type can be obtained in different ways. In Figure 10.4, it is determined by assuming that the unit cost to date will continue to completion of that work type. Multiplying the total estimated quantity by the underrun or overrun of the unit price to date yields the projected saving or loss figure. Another way the projected values can be computed is through the use of unit-price trends. Recent unit costs can be used to forecast final cost variances.

The projected saving and loss figures in columns 15 and 16 of Figure 10.4 afford a quick, informative summary of how the project is doing insofar as labor cost is concerned. Those work types with labor overruns are identified together with the

WEEKLY LABOR COST REPORT

Project: Highway Bridge
Week ending: July 21 (working day 27)

Project No. 9108-05
Prepared by R.H.C.

Cost Code (1)	Work Type (2)	Unit (3)	Quantity			Direct Labor Cost			Labor Unit Cost			To Date		Projected	
			Budget (4)	This Week (5)	To Date (6)	Budget (7)	This Week (8)	To Date (9)	Budget (10)	This Week (11)	To Date (12)	Savings (13)	Loss (14)	Savings (15)	Loss (16)
02220.10.3	Excavation, unclassified	cy	1,667	0	1,667	$1,308	$0	$1,217	$0.78	$0.00	$0.73	$91		$91	
02222.10.3	Excavation, structural	cy	120	0	120	$1,520	$0	$1,686	$12.67	$0.00	$14.05		$166		$166
02350.00.3	Piledriving, rig mob. & demob.	job	–	–	–	*$2,337	$0	$2,140	$2,337.00	$0.00	$2,140.00	$197		$197	
02361.10.3	Piling, steel, driving	lf	2,240	560	2,016	$4,345	$5,431	$8,328	$1.94	$9.70	$4.13		$4,417		$6,155
03150.10.3	Footing forms, fabricate	sf	360	0	360	$666	$0	$685	$1.85	$0.00	$1.90		$19		$19
03150.20.3	Abutment forms, prefabricate	sf	1,810	0	1,810	$1,991	$0	$1,560	$1.10	$0.00	$0.86	$431		$431	
03157.10.3	Footing forms, place	sf	720	0	360	$242	$0	$111	$0.34	$0.00	$0.31	$11		$22	
03159.10.3	Footing forms, strip	sf	720	360	360	$104	$48	$48	$0.14	$0.13	$0.13	$4		$8	
03157.20.3	Abutment forms, place	sf	3,620	1,810	1,810	$5,853	$2,810	$2,810	$1.62	$1.55	$1.55	$116		$233	
03311.10.3	Concrete, footing, place	cy	120	0	60	$611	$0	$282	$5.09	$0.00	$4.70	$23		$47	
											Subtotals	$873	$4,602	$1,029	$6,340
											Totals		$3,729		$5,311

*67% of budget for mobilization

Figure 10.4 *Highway bridge, weekly labor cost report (#1).*

WEEKLY LABOR COST REPORT

Project __Highway Bridge__   Project No. __9108-05__
Week ending __July 21 (working day 27)__   Prepared by __R.H.C.__

Cost Code (1)	Work Type (2)	Unit (3)	Total Quantity Budgeted (4)	Total Quantity to Date (5)	Percent Com- pleted (6)	Budgeted Direct Labor (7)	Budgeted Labor to Date (8)	Actual Labor to Date (9)	Cost Difference (10)	Deviation (11)
02220.10.3	Excavation, unclassified	cy	1,667	1,667	100%	$1,308	$1,308	$1,217	$91	0.93
02222.10.3	Excavation, structural	cy	120	120	100%	$1,520	$1,520	$1,686	-$166	1.11
02350.00.3	Piledriving, rig mob. & demob.	job	—	—	100%	*$2,337	$2,337	$2,140	$197	0.92
02361.10.3	Piling, steel, driving	lf	2,240	2,016	90%	$4,345	$3,911	$8,328	-$4,418	2.13
03150.10.3	Footing forms, fabricate	sf	360	360	100%	$666	$666	$685	-$19	1.03
03150.20.3	Abutment forms, prefabricate	sf	1,810	1,810	100%	$1,991	$1,991	$1,560	$431	0.78
03157.10.3	Footing forms, place	sf	720	360	50%	$242	$121	$111	$10	0.92
03159.10.3	Footing forms, strip	sf	720	360	50%	$104	$52	$48	$4	0.92
03157.20.3	Abutment forms, place	sf	3,620	1,810	50%	$5,853	$2,927	$2,810	$117	0.96
03311.10.3	Concrete, footing, place	cy	120	60	50%	$611	$306	$282	$24	0.92
Totals to Date							$15,138	$18,867	-$3,730	1.25

*67% of budget for mobilization

Figure 10.5 Highway bridge, weekly labor cost report (#2).

financial consequences if nothing changes. To illustrate, the effect of the underground boulders on the pile driving for abutment #2 (discussed in Section 9.11) is now obvious and probably will account for a direct labor cost overrun of more than $6,000. Some labor cost reports indicate the trend for each cost code, that is, whether the unit cost involved has been increasing or decreasing. This information can be helpful in assessing whether a given cost overrun is improving or worsening and in evaluating the efficacy of cost reduction efforts.

Figure 10.5 is an alternative form of weekly labor cost report that shows actual and budgeted labor costs to date for each cost classification. In this figure, the budgeted total quantities and budgeted total labor cost for each cost code are obtained from the project budget, Figure 3.10. The actual work quantities and labor costs to date are cumulative totals for each work classification obtained from the time cards and weekly quantity reports. Column 10 of Figure 10.5 shows the cost difference as column 8 minus column 9, with a positive difference indicating that the cost as estimated exceeds the actual cost to date. Hence in column 10, a positive number is desirable; a negative number, undesirable. The deviation is the actual cost to date (column 9) divided by the estimated cost to date (column 8). A deviation of less than one indicates that labor costs are within the budget, whereas a deviation of more than one indicates a cost overrun.

Although column 10 does indicate the magnitude of the labor cost variation for each cost code, it does not indicate the relative seriousness of the cost overruns. The deviation is of value in this regard because it shows the relative magnitude of labor cost variance. For those work types not yet completed, the cost differences listed in column 10 of Figure 10.5 do not always check exactly with the to-date saving and loss values in Figure 10.4. These small variations are caused by the rounding off of numbers and are not important.

10.16 EQUIPMENT COST ACCOUNTING

When equipment costs constitute a substantial portion of the cost of construction, the determination and analysis of field equipment expense are an important part of project cost accounting. Equipment accounts for a substantial proportion of the cost of highway, heavy, and utility construction, and the control of equipment expense on such projects is as important as the control of labor costs. Equipment expense, like labor cost, is inherently uncertain and subject to unpredictable variations. The management and control of equipment expense require a comprehensive equipment cost-accounting system. The objectives of equipment cost accounting are the same as those discussed for labor costs. Management requires timely cost and production information for effective project cost control, and estimators need equipment productivity data to estimate the costs of future work. Only major equipment items require detailed cost study, however. Less expensive types such as power saws, concrete vibrators, and hand-operated soil compactors do not receive or merit detailed cost analysis.

During the field construction period, the equipment cost-accounting system determines how many hours each equipment unit is in operation and the project work accounts to which these hours apply. Using equipment hourly rates, equipment costs are determined periodically for each work type. These costs are matched with work quantities produced, and equipment unit prices are computed. The actual equipment costs are compared with those budgeted in weekly equipment cost reports. The equipment time rates of expense charged against the project are those used when the job was initially cost estimated (the determination of these equipment budget rates was discussed in Section 3.18). Equipment cost accounting is, in most respects, similar to that for labor.

10.17   CHARGING EQUIPMENT TO THE PROJECT

The accounting procedures used to charge construction projects with the ownership and operating costs of equipment vary substantially from one contractor to another. On projects that have relatively modest equipment requirements, contractors will frequently charge all equipment costs to a single equipment cost account and make no attempt to distribute these expenses to the work types on which the equipment was actually used. This procedure is not suitable or adequate, however, for projects where large spreads of equipment are involved.

A more common approach to charging equipment costs to projects is to assign equipment expense to the appropriate work accounts just as is done with labor. During the progress of the work in the field, a charge equal to the budget rate times the number of equipment hours, weeks, or months worked is made against the cost accounts involved. At the same time, an equal, but opposite, credit is made to the account of that equipment unit. This is the same equipment account previously discussed in Section 3.18. In essence, what the contractor is doing is establishing, on paper only, a separate company that owns, services, and maintains all of the major equipment and rents it to the contractor at predetermined rates. Commensurately, all equipment costs, exclusive of labor, are charged to this fictitious company through the medium of ascribing them to the individual equipment accounts. The contractor assesses the cost of equipment usage to its job by charging the individual project work accounts at the established internal rental rates. These same equipment charges are credited to each equipment account as a rental payment.

This equipment accounting procedure provides a cumulative record of expense and earnings for each major equipment unit. How the figures compare provides company management with invaluable information concerning equipment use, maintenance, costs, and replacement. In short, the company knows which machines are paying their way and which are not. Many construction firms have a company policy that when the annual costs of an equipment item exceed its annual earnings, it is either replaced or sold.

The internal rental rates used to charge equipment time to projects are based on time-average ownership and operating expenses that actually vary over the service life of the equipment. To illustrate, investment costs decrease and repair costs increase with equipment age. However, the use of lifetime average costs is the only way to have each project bear its proper share of the ultimate total expense associated with any particular equipment item. When equipment budget rates are assessed to a job, this is an all-inclusive charge. Correspondingly, the costs of fuel, lubrication, maintenance, repairs, and other such equipment expenses are not charged to the job on which they are actually incurred, but to the applicable equipment accounts.

10.18 EQUIPMENT TIME REPORTS

Because equipment costs are expressed as a time rate of expense, time reporting is the starting point for equipment cost accounting. Equipment time is kept in much the same way as labor time. Where major equipment items are involved, a common procedure is to have the equipment supervisor make out daily or weekly equipment time cards that include each equipment item on the job. Equipment time cards are separate from and in addition to the operators' time cards. This procedure has merit because, by using different time cards, separate reportings are available for payroll and labor cost-accounting purposes and for equipment cost accounting. In addition, equipment items such as pumps and air compressors may not have full-time operators and could be overlooked otherwise. Figure 10.6 is a typical daily equipment time card.

The equipment time card performs the same cost accounting function as the labor time card. By allocating equipment times to the proper cost codes, it is possible to determine the equipment costs chargeable to the various work types. Accuracy of time allocation to cost codes is just as important for equipment as it is for labor if reliable information is to be obtained for purposes of cost control and estimating. Even if a weekly time card is used, it is preferable that the time information be entered on a daily basis. Just as with labor time, distributing the equipment time daily is conducive to better accuracy. Someone other than the field supervisor, such as the cost engineer, may fill out the time card and enter the budget rates of the individual equipment items reported and make the cost extensions.

Figure 10.6 records equipment time as working time (W), repair time (R), and idle time (I). Excessive equipment idle time (it may be difficult to get this item reported honestly) may indicate field management problems such as too much equipment on the job, lack of operator skill, improper balance of the equipment spread, or poor field supervision. Appreciable repair time can indicate inadequate equipment maintenance, worn-out equipment, severe working conditions, or operator abuse. On the other hand, substantial unproductive time can also be caused by job accidents, inclement weather, unanticipated job problems, and unfavorable site conditions.

10.19 WEEKLY EQUIPMENT COST REPORT

Once each week, equipment costs are matched with the corresponding quantities of work produced. Work quantities are derived from the weekly quantity report previously dealt with in Sections 10.11 and 10.12. By following the same process as has already been described for labor, a weekly equipment cost report is prepared. An equipment cost report summarizes all equipment costs incurred on the project up through the effective date of the report. Either of the two cost report forms previously used for labor (Figures 10.4 and 10.5) can be used for equipment. Figure 10.7 is a

DAILY EQUIPMENT TIME CARD

Project __Highway Bridge__   Project No. ___9108.05___   Weather __Cloudy-windy__

Date __July 20__   Prepared by __R.H.C.__

Equipment No.	Equipment Type	Operator	Hourly Rate	Cost Code		Total Hours			Total Cost
				02361.10.4	03157.20.4	W	R	I	
411	55 ton crane	Mills	$108.00	8		8	0	0	$864.00
235	600 cfm compressor	—	$12.38	6	1	7	1	0	$86.66
662	1-cy backhoe	Milne	$26.00	8		6	0	2	$208.00
Total Cost				$1,146.28	$12.38				$1,158.66

Figure 10.6 Highway bridge, daily equipment time card.

frequently used format for weekly equipment cost reports. As can be seen, this figure is very similar to the weekly labor cost report form shown in Figure 10.4. Figure 10.7 serves to inform project management in a quick and concise manner how equipment costs are faring and about those work items where costs are going over the budget. The equipment report form in Figure 10.7 presents both work quantities and equipment expense and yields actual unit costs for each work type. Comparing the estimated with the actual equipment unit costs discloses where equipment costs are overrunning the project budget. The budgeted quantity, budgeted equipment cost, and budgeted equipment unit cost for each work type are taken from the project budget, Figure 3.10. The other quantities and equipment costs are actual values, either for the week reported or to date.

In Figure 10.7, the to-date and projected saving and loss values capsulize the equipment cost experience on the highway bridge up through July 21. Of the work types that have been accomplished or that are presently in progress, only the equipment expense involved with the driving of the steel pilings gives serious concern. As of July 21, this item of work was 90 percent completed, and its equipment cost had overrun the budget by $4,689. There is every indication that this work category will incur an equipment expense loss of about $6,438 by the time it is finally completed.

10.20 SPECIAL ASPECTS OF EQUIPMENT CHARGES

The usual procedure for charging equipment ownership and operating costs to construction projects was discussed in Section 10.17. There are, however, some aspects of equipment charges that require special arrangements. For example, some equipment expenses are not included in the usual hourly budget rates. The costs of move-in, erection, dismantling, and move-out are fixed costs that cannot be incorporated into time rates of expense. Such costs are normally charged to appropriate job overhead accounts and are not included in the equipment hourly or monthly rates. With regard to equipment charged to the project at an hourly rate, the handling of idle and repair time is a matter of company policy. Several different procedures are followed in this regard. Probably the most common approach is to charge the project at the established internal rates for the full working day for each piece of equipment on the job. However, credit is given for repair time and for idle time caused by weather and other uncontrollable causes. The significance of this procedure is that the job is charged for all equipment on the site whether it is used or not. This policy can materially assist in controlling underusage of equipment. Where standby equipment units are purposely kept on a job to handle emergencies, the project is often charged only with the ownership expense involved.

Where the project is charged with extensive periods of equipment idle time, it is probably best to charge the individual cost accounts with net operating hours, plus ordinary or usual idle time. The excessive idle time can be charged to a special overhead account. The reason for this accounting maneuver is that the equipment cost charged to a given work item is used to compute summary information for purposes of cost control and estimating. The use of the special account avoids having excessive idle time distort the reported equipment unit costs.

WEEKLY EQUIPMENT COST REPORT

Project Highway Bridge
Week ending July 21 (working day 27)

Project No. 9108-05
Prepared by R.H.C.

Cost Code (1)	Work Type (2)	Unit (3)	Quantity			Direct Equipment Cost			Equipment Unit Cost			To Date		Projected	
			Budget (4)	This Week (5)	To Date (6)	Budget (7)	This Week (8)	To Date (9)	Budget (10)	This Week (11)	To Date (12)	Savings (13)	Loss (14)	Savings (15)	Loss (16)
02220.10.4	Excavation, unclassified	cy	1,667	0	1,667	$619	$0	$580	$0.37	$0.00	$0.35	$39		$39	
02222.10.4	Excavation, structural	cy	120	0	120	$390	$0	$420	$3.25	$0.00	$3.50		$30		$30
02350.00.4	Piledriving, rig mob. & demob.	job	—	—	—	*$3,672	$0	$3,420	$3,672.00	$0.00	$3,420.00	$252		$252	
02361.10.4	Piling, steel, driving	lf	2,240	560	2,016	$5,676	$5,790	$10,365	$2.53	$10.34	$5.14		$4,689		$6,438
03157.20.4	Abutment forms, place	sf	3,620	1,810	1,810	$1,814	$875	$875	$0.50	$0.48	$0.48	$32		$64	
03311.10.4	Concrete, footing, place	cy	120	0	60	$580	$275	$275	$4.83	$0.00	$4.58	$15		$30	
											Subtotals	$338	$4,719	$346	$6,507
											Totals		$4,381		$6,161

*67% of budget for mobilization

Figure 10.7 Highway bridge, weekly equipment cost report.

10.21 MONTHLY COST FORECAST

The discussions so far in this chapter have described how weekly labor and equipment cost reports are generated and the information they contain. Material and subcontract expenses are also incurred on the project, but these two cost categories are not as volatile as labor and equipment and do not require constant and detailed monitoring. Nevertheless, all project costs must be summarized and reported at regular intervals, monthly being common. The report used for this purpose also forecasts the final total job cost. Figure 10.8 presents a form of cost forecast report used to inform project management each month concerning the overall financial position of the job. Unlike the weekly labor cost reports, the labor cost figures reported in the monthly cost forecast include both direct and indirect labor expense. This is necessary, of course, to present job management with a complete review of total job costs.

To prepare a cost forecast report, the information shown in Figure 10.8 pertaining to materials, labor, equipment, and subcontracts are first computed for each elementary work type on the job. These values are then combined and costs are presented only for the major work classifications as is indicated by Figure 10.8. Summary costs are listed for materials, labor, equipment, and subcontracts. Using labor as an example, column 7 lists the total budgeted labor cost for each major cost code listed. Following this is the total labor cost to date and the estimated total labor cost to complete each work account. The variance, column 10, is obtained by subtracting the budgeted cost (column 7) from the sum of cost to date (column 8) and estimated cost to complete (column 9). Variance is thus the difference between anticipated actual cost and budget. A positive value of variance is the amount by which the actual cost is expected to exceed the estimated cost. A negative variance is, of course, just the opposite. The costs shown under materials, equipment, and subcontracts are obtained in the same fashion.

Column 19 in Figure 10.8 is obtained as the sum of the labor, material, equipment, and subcontract budgeted costs. Columns 20 and 21 are similarly calculated. The estimated total final cost (column 22) for each cost code is obtained by adding the total cost to date (column 20) to the total estimated cost to complete (column 21). The variance (column 23) is the total budgeted cost (column 19) subtracted from the estimated total final cost (column 22). The percent completion (column 24) for each major cost code is obtained by dividing the total cost to date (column 20) by the estimated total final cost (column 22). The algebraic sum of the total variance values (column 23) is the amount by which it now appears the actual cost will be less than or more than the budgeted cost for the total project. In Figure 10.8, it appears that, as of July 30, the actual total cost of constructing the highway bridge will be $7,405 more than had been originally estimated.

For completed items, the actual final cost is used for the estimated total final cost. The usual assumption for work underway is that it will be finished at its current cost rate and that unstarted work will be completed as budgeted. Exceptions to this would be when better cost information is now available or reduced unit costs reflecting learning curve effects are in order.

MONTHLY COST FORECAST REPORT

Project __Highway Bridge__

Period Ending __July 31__

Cost Code (1)	Description (2)	Materials				Labor				Equipment			
		Budget (3)	Cost to Date (4)	Est. to Complete (5)	Variance (6)	Budget (7)	Cost to Date (8)	Est. to Complete (9)	Variance (10)	Budget (11)	Cost to Date (12)	Est. to Complete (13)	Variance (14)
01	General Requirements	$0	$0	$0	$0	$6,467	$6,220	$0	-$247	$4,038	$3,760	$0	-$278
02	Sitework	$30,690	$30,490	$200	$0	$12,311	$15,433	$1,632	$4,754	$12,662	$11,731	$5,900	$4,969
03	Concrete	$27,910	$16,004	$12,150	$244	$24,130	$9,232	$12,795	-$2,103	$7,645	$3,341	$4,100	-$204
05	Metals	$33,040	$0	$33,040	$0	$3,404	$0	$3,404	$0	$2,592	$0	$2,592	$0
06	Finishes	$0	$0	$0	$0	$0	$0	$0	$0	$0	$0	$0	$0

Figure 10.8   Highway bridge, monthly cost forecast report.

Project No.   9108.05

Prepared by   R.H.C.

Cost Code (1)	Subcontracts				Totals					
	Budget (15)	Cost to Date (16)	Est. to Complete (17)	Variance (18)	Budget (19)	Cost to Date (20)	Est. to Complete (21)	Estimated Final Cost (22)	Estimated Variance (23)	Percent Complete (24)
01	$0	$0	$0	$0	$10,505	$9,980	$0	$9,980	-$525	100%
02	$0	$0	$0	$0	$55,663	$57,654	$7,732	$65,386	$9,723	88%
03	$0	$0	$0	$0	$59,685	$28,577	$29,045	$57,622	-$2,063	50%
05	$40,275	$21,748	$18,797	$270	$79,311	$21,748	$57,833	$79,581	$270	27%
06	$5,820	$0	$5,820	$0	$5,820	$0	$5,820	$5,820	$0	0%
					$210,984	$117,959	$100,430	$218,389	$7,405	54%

Figure 10.8 continued.

10.22   TIME-COST ENVELOPE

In this chapter, methods have been discussed that enable project management to check actual field costs against an established budget. The principal emphasis has been to provide detailed cost information and forecasts for use by field supervisors. There is also a need for a more general kind of cost report that provides a quick and concise picture of the overall cost status of the project. This summary information is for use by top project management, owners, lending agencies, and others. The monthly cost forecast report of the previous section can serve in this regard. Project time-cost envelopes are graphical cost summaries that are also used for job cost monitoring.

A cost envelope is obtained by plotting cumulative estimated job cost against construction time. Two such curves are drawn, one on the basis of an early start schedule and the other on a late start basis. This concept has been discussed previously in Section 9.19. Figure 10.9 shows a typical project time-cost envelope.

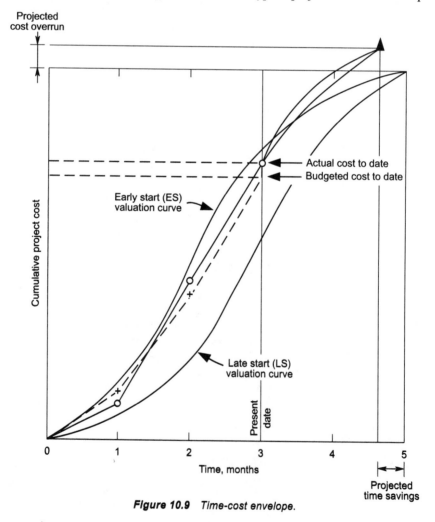

Figure 10.9   *Time-cost envelope.*

Each month the total actual cost to date is plotted. These values are shown in Figure 10.9 as small circles. Alone, this information is inconclusive, but when the budgeted cost of the work accomplished to date is plotted concurrently, several aspects of the project cost and time status emerge. The budgeted costs to date are shown as crosses in Figure 10.9.

To illustrate how the cost information in Figure 10.9 is interpreted, note the project cost status as of the end of the third month. This is indicated as "present date" in the figure. The actual cost to date plots above the budgeted cost to date indicating a cost overrun for the project as a whole. The same situation existed at the end of the previous month and the cost overrun is increasing. The budgeted cost to date appears near the early start curve of the cost envelope, indicating that the project is well within the schedule time limits.

Based on the monthly cost forecast report as of the end of the third month, a new cost envelope can be generated from "present date" through the end of the job. This projection in Figure 10.9 shows that there is a projected project cost overrun and a projected early completion. Time-cost envelopes such as shown in Figure 10.9 are excellent for graphically displaying time and cost trends for the total project.

10.23 SPECIAL COST-ACCOUNTING PROBLEMS

There are numerous practical considerations that give rise to special problems concerning the proper operation of a project cost-accounting system. Satisfactory solution of these problems can be fundamental to the success or failure of a project cost system. It follows, therefore, that project management must be aware of such difficulties and take appropriate steps to solve them. What follows is meant merely to afford some specific examples and not to be an exhaustive review of all such cases.

Usually, equipment expense is directly chargeable to a single project work account. However, there are frequently occasions where this is not true, and equipment costs must be accumulated in "suspense" or "clearing" accounts until such time as they can be distributed to the proper cost accounts. Many examples of this could be cited, but one might be a central concrete-mixing plant, consisting of many separate equipment units, that is, producing concrete for several different cost accounts. Expenses for equipment of this type can be collected into a suspense account and be periodically distributed equitably to the appropriate cost accounts on the basis of quantities involved. Thus, each cost account is charged in the same manner as if the concrete had been purchased from an outside company. All suspense account costs must ultimately be charged to the end-use accounts leaving a zero balance.

A different case involves an equipment item that serves many different operations. A tower crane on a high-rise building might be such an instance. Over a short period of time, this crane could handle structural steel, concrete, masonry, reinforcing steel, door frames, electrical conduit, pipe, and any number of other materials. Allocation of this crane's time to specific cost codes would be an all but impossible task. A common approach is to establish a special cost category for the crane in project overhead. All crane costs are charged to this account with no effort made to distribute its time to specific work classifications.

Accounting for bulk materials is another common example of where clearing accounts are used. Stock items such as lumber, pipe, and unfabricated reinforcing steel may ultimately be used in conjunction with many different cost accounts. In such cases, a clearing account is created to accumulate all costs associated with the materials until they are put to use. As they are used, the suspense account is credited and the appropriate work accounts are charged.

A different kind of cost accounting problem is encountered when an owner imposes on the contractor a cost code different from the contractor's own system. Owners sometimes have need for particular cost information for purposes of capital depreciation, taxation, and insurance. Some owners, such as public utilities, may require cost values that fit standard account classifications imposed by regulatory agencies. Construction contracts of this kind can require the contractor to maintain project costs in two or even three different coding systems. Conversion from one system to another can normally be automated by computer, however.

10.24   PRODUCTION COST REDUCTION

The project cost reports generated by the cost-accounting system make it possible for job management to assess the cost status of the work and to pinpoint those work areas where expenses are excessive. In this way, management attention is quickly focused on those work classifications that need it. If the project expense information is developed promptly, it may be possible to bring the offending costs back into line. In actual fact, of course, project cost control starts when the job is first priced because this is when the control budget is established. No amount of management expertise or corrective action can salvage a project that was priced too low in the beginning. In this regard, there likely will always be some work classifications whose actual costs will exceed those estimated. The project manager is primarily responsible for getting the total project built for the estimated cost. If some costs go over, hopefully they can be counterbalanced by savings in other areas.

Having identified where production costs are excessive, project management must then decide just what to do about them, if indeed anything can be done. For certain, the hourly rates for labor and equipment are not controllable by management. The only real opportunity for cost control resides in the area of improving production rates. This element of work performance can, to a degree, be favorably influenced by skilled field supervision, astute job management, energetic resource expediting, and the improved makeup of labor crews and selection of equipment.

Any efforts to improve field production must be based on detailed knowledge of the pertinent facts. If the cause of excessive costs is not specifically identified, then a satisfactory solution is not likely to be found. It is impossible to generalize on this particular matter, but certainly the treatment must be gauged to the disease. In the usual case, full cooperation between the field supervisors and project management is needed before any real cost improvement can be realized. Field supervisors play a key role in implementing corrective procedures. There are no precise guidelines for reducing excessive project costs. The effectiveness of corrective procedures depends largely on the ingenuity, resourcefulness, and energy of the people involved.

Production costs are frequently too high early in the construction process, but tend to become lower as the work progresses. For instance, the unit costs of forming abutment #1 on the highway bridge would be expected to be higher than those for the second one. One reason for this is that, on repetitive operations, costs have a tendency to decline as experience increases. Also, some work done on the first abutment such as cutting wales to length, cutting and splicing braces, and drilling holes for form ties will not have to be repeated for the second abutment. Production rates generally have a tendency to increase as the job proceeds because tradesmen become familiar with the job and learn how to work with one another as a team.

10.25   INFORMATION FOR ESTIMATING

Estimating needs production rates and unit costs that are a balanced time average of good days and bad days, high production and low production. For this reason, information for estimating is normally not recovered from the cost-accounting system until after project completion, or at least not until all the work type being reported has been finished. In so doing, the best possible time-average rates will be obtained. Permanent files of cost and productivity information are maintained, providing the estimator with immediate access to data accumulated from prior projects.

Both production rates and unit costs are available from the project cost-accounting system. To be of maximum value in the future, however, it is important that such productivity data be accompanied by a description of the project work conditions that applied while the work was being done. Knowledge of the work methods, equipment types, weather, problems, and other job circumstances will make the basic cost and productivity information much more useful to an estimator. Such written narrative becomes a part of the total historical record of each cost account.

10.26   COMPUTER APPLICATION

Computers are universally used by construction contractors in conjunction with their project cost systems. Because cost accounting and analysis can become laborious and time-consuming, even for relatively small operations, the computer has economic, speed, and accuracy advantages over manual methods. In addition, the computer provides a cost system with flexibility and depth that manual systems often cannot match. This does not mean that job costs cannot be developed satisfactorily by hand. Many small contractors have perfectly adequate manual cost systems. Experience indicates, however, that few contractors of substantial size are able to manually generate field cost reports in time to serve a genuine cost-control purpose. Contractors often find that manual methods serve well as a generator of estimating information, but not for cost control. A common experience in this regard is that the project is finished before the contractor knows the profit status of the work. It is only being realistic to recognize that most contractors find computer support to be a necessary part of their project cost system.

Supporting computer software is now very flexible in the sense that just about any cost report or information can be generated that project management may desire. Cost reports in the same general formats as those presented in this chapter are produced by several current computer programs. The programs commonly used by contractors actually perform a whole series of cost-accounting and financial accounting functions. After input of cost and production information, the computer generates payroll checks, keeps payroll records, maintains the equipment accounts, and performs other functions as well as producing a variety of productivity and cost reports and project cost forecasts.

10.27 ACCURACY OF ESTIMATING

A functioning project cost system acts as a detailed check on estimating accuracy. However, many contractors do not consistently maintain a viable project cost system. In addition, the essential thrust of a cost-accounting system is not directed toward evaluating the overall "accuracy" of the estimating process itself, with accuracy here referring to the magnitude of the deviation between the total estimated cost of a project and its total actual cost. Accuracy in the computation of construction costs is crucial to the success of a construction firm, and studies of estimating performance are an important element in making periodic reviews of company performance. An occasional analysis of estimating accuracy will provide a construction company with revealing information of great value in assessing the efficacy of its estimating procedures and practices.

In making a study of estimating accuracy, some simple statistical methods can be of great assistance in obtaining summary data and interpreting the results. A discussion of these procedures and a numerical example are presented in Appendix E.

11

PROJECT FINANCIAL MANAGEMENT

11.1 FINANCIAL CONTROL

The previous chapters have discussed the essential elements of construction time, cost, and resource control. There is one additional major feature of the project management system that remains to be treated: financial control. The project manager bears the overall responsibility for the financial management of the work. This includes carrying out such fiscal duties as may be imposed by the construction contract and implementing appropriate monetary procedures according to the dictates of good business practice. Project financial management can involve a broad range of responsibilities.

Construction contracts normally require that contractors perform prescribed duties of a financial nature. For example, they are made responsible for certain aspects of the payment process. This can include project cost breakdowns, forecasted schedule of progress payments, preparation or approval of periodic pay estimates, and documentation required for final payment. Construction contracts prescribe specific procedures to be followed by the contractor with regard to payment for extra work, extensions of time, processing of change orders, claims, and settlement of disputes.

The project manager is also responsible to his company for implementing and maintaining standard fiscal procedures. One of the most important of these is monitoring the project cash requirements during the contract period. Even a highly profitable job will require a considerable amount of cash to meet payrolls, purchase materials, and meet other project obligations. The size and timing of these cash demands is a serious matter for the contractor and appropriate financial forecasts must be made. A system of disbursement control is needed to regulate and control payments to material vendors, subcontractors, and others.

Another aspect of financial control is that of maintaining a complete and detailed daily record of the project. Such a job log can be invaluable in the settlement of claims and disputes that may arise from the work. Such a job history includes names, dates, places, and documentation of everything that happens as well as everything that fails to happen.

11.2 PROGRESS PAYMENTS

Construction contracts typically provide that the owner shall make partial payments of the contract amount to the prime contractor as the work progresses. Payment at monthly intervals is the usual proviso. Depending on the type of work and contract provisions, the monthly pay requests may be prepared by the contractor, the architect–engineer, or the owner. In any event, a pay request is prepared periodically that compiles the cost of the work accomplished since the last payment to the contractor was made by the owner. This is typically done in practice by determining the total value of work actually performed to date and then subtracting the sum of the previous progress payments made by the owner.

The total value of work done to date is obtained in different ways, depending on the type of contract. Under lump-sum contracts, progress is customarily measured in terms of estimated percentages of completion of major job components. The quantities of work done on unit-price contracts are determined by actual field measurement of the bid items put into place. In either type of contract, materials stored on site are customarily taken into account as well as any prefabrication or preassembly work that the contractor may have done at some location other than the job site.

A prescribed percentage of each progress payment is usually retained by the owner in accordance with the terms of the contract. A retainage of 10 percent is common, although other percentages are also used. To an increasing extent, construction contracts provide that retainage shall be withheld only during the first half of the project. After that, if the work is progressing satisfactorily and with the consent of the surety, all subsequent progress payments are made in full. The retainage is held by the owner until the work receives final certification by the architect-engineer, the owner accepts the project, and the contractor submits any required affidavits and releases of lien. Final payment is then made to the contractor, including the accumulated retainage.

Negotiated contracts of the cost-plus variety usually provide for the contractor's submission of payment vouchers to the owner at specified intervals during the life of the contract. A common provision is weekly reimbursement of payrolls and monthly reimbursement of all other costs, including a pro rata share of the contractor's fee. It is not uncommon under this type of contract for the owner to pay all vouchers in full without deducting any percentage as retainage. Some contracts provide for the retention of a stated percentage of the contractor's fee. Others provide that the owner make full reimbursement to the contractor up to some designated percentage (80 percent is sometimes used) of the total estimated project cost. Further payments are then withheld until some specified amount of money has been set aside. This reserve is retained by the owner until the project has been satisfactorily completed.

11.3 PAY REQUESTS FOR UNIT-PRICE CONTRACTS

Payment requests under unit-price contracts are based on actual quantities of each bid item completed to date. This is the case with the highway bridge that was bid, as shown in Figure 3.9, as 12 different bid items. The measurement of quantities in the field is done in several different ways depending on the nature of the particular bid item. When cubic yards of aggregate, tons of asphaltic concrete, or bags of portland cement are set up as bid items, these quantities are usually measured as they are delivered to the work site. Delivery tickets or fabricator's certificates are used to establish tons of reinforcing or structural steel. Other work classifications such as cubic yards of excavation, lineal feet of pipe, or cubic yards of concrete are measured in place or computed from field dimensions. Survey crews of the owner and of the contractor often make their measurements independently and adjust any differences.

Many owners use their own standard forms for monthly pay estimates. On unit-price contracts, the owner often prepares the pay request and sends it to the contractor for checking and approval before payment is made. On unit-price contracts that involve a substantial number of bid items, each monthly pay request is a sizable document consisting of many pages. In essence, the total amount of work accomplished to date of each bid item is multiplied by its corresponding contract unit price. All of the bid items are totaled and the value of materials stored on the site is then added. From this total is subtracted the prescribed retainage. The resulting figure represents the entire amount due the contractor for its work to date. The sum of all prior progress payments that have already been paid is then subtracted, this yielding the net amount of money payable to the contractor for its work that month. Figure 11.1 is the pay request for the highway bridge covering the work accomplished during the month of July.

As explained, monthly pay requests on unit-price contracts are based on the bid-item quantities actually accomplished. One way work quantities could be obtained is from progress measurements on network activities. However, while activities can serve well as a means of obtaining work quantities put into place for cost-accounting purposes, they do not usually constitute an acceptable basis for compiling monthly pay requests on unit-price projects. Unit-price contracts often do not have well-defined quantities shown on the construction drawings. Activity quantities taken from the construction drawings before work commences may not be accurate representations of the work amounts actually accomplished. Under unit-price contracts, payment to the contractor must be based on field-determined quantities of the individual bid items. Even though activity quantities work well to establish cost control on a weekly basis, they much occasionally be checked and corrected by actual measurement. The field quantity measurements made monthly for payment purposes also accomplish this purpose.

PAYMENT REQUEST

Project Highway Bridge
Periodic Payment No. 2 for Period   July 1 - July 31

Item No.	Description	Quantity	Unit	Unit Price	Amount
1	Excavation, unclassified	1,667	cy	$2.10	$3,500.70
2	Excavation, structural	120	cy	$29.53	$3,543.60
3	Backfill, compacted	102	cy	$11.40	$1,162.80
4	Piling, steel	2,240	lf	$34.02	$76,204.80
5	Concrete, footings	120	cy	$113.62	$13,634.40
6	Concrete, abutments	140	cy	$240.65	$33,691.00
7	Concrete, deck slab, 10 in.	0	sy	$99.03	$0.00
8	Steel, reinforcing	36,817	lb	$0.65	$23,931.05
9	Steel, structural	0	lb	$0.69	$0.00
10	Bearing plates	0	lb	$1.66	$0.00
11	Guardrail	0	ft	$65.95	$0.00
12	Paint	0	ls	$8,394.55	$0.00

Construction performed to date	$155,668.35
Materials stored on site (schedule attached)	$0.00
Total work performed and materials stored	$155,668.35
Less 10% retention	$15,566.84
Net work performed and materials stored	$140,101.52
Less amount of previous payments	$790.00
Balance due this payment	$139,311.52

Figure 11.1   Highway bridge, monthly pay request.

11.4   PROJECT COST BREAKDOWN

Construction contracts for lump-sum projects normally require that a cost breakdown of the project be submitted by the contractor to the owner or architect–engineer for approval before submittal of the first pay request. This cost breakdown, also called a "schedule of values," serves as the basis for subsequent monthly pay requests. The breakdown, which is actually a schedule of costs of the various components of the structure, is prepared to assist the owner or architect–engineer in checking the contractor's pay requests. It is to be noted here that all of the example projects described in this text have been unit price. Consequently, at this point it is necessary for illustrative purposes to introduce data for a lump-sum job taken from an external source. Figure 11.2, which presents a typical monthly pay request for a lump-sum job, shows the cost breakdown of the project in column 1. Occasionally, the owner will specify the individual work items for which the contractor is to present cost figures. In the absence of such instructions, it is usual to prepare the cost schedule using the same general items as they appear in the specifications and on the final recap sheet of the estimate. This practice minimizes the time and effort required to compute the breakdown values and gives a maximum of accuracy in the results.

THE BLANK CONSTRUCTION COMPANY, INC.
1938 Cranbrook Lane
Portland, Ohio

PERIODIC ESTIMATE FOR PARTIAL PAYMENT

Project Municipal Airport Terminal Building   Location Portland, Ohio

Periodic Estimate No.   4   For Period   September 1, 19-- to September 30, 19--

Item No.	Item Description	Total Cost (1)	Completed to Date (2)	Cost to Complete (3)	Percent Complete (4)
1	Clearing and Grubbing	$14,909	$14,909	$0	100
2	Excavation and Fill	$44,749	$38,037	$6,712	85
3	Concrete and Forms				
	Footings`	$71,915	$51,060	$20,855	71
	Grade Beams	$72,131	$41,836	$30,295	58
	Beams	$43,690	$4,369	$39,321	10
	Columns	$14,113	$4,234	$9,879	30
	Slabs	$253,749	$30,450	$223,299	12
	Walls	$39,455	$14,204	$25,251	36
	Stairs	$28,734	$0	$28,734	0
	Sidewalks	$23,622	$0	$23,622	0
4	Masonry	$486,566	$14,597	$471,969	3
5	Carpentry	$37,772	$0	$37,772	0
6	Millwork	$85,634	$0	$85,634	0
7	Steel and Misc. Iron				
	Reinforcing Steel	$85,590	$44,507	$41,083	52
	Mesh	$18,800	$2,820	$15,980	15
	Joist	$92,953	$0	$92,953	0
	Structural	$178,779	$30,392	$148,387	17
8	Insulation	$25,688	$0	$25,688	0
9	Caulk and Weatherstrip	$5,741	$0	$5,741	0
10	Lath, Planter, and Stucco	$196,240	$0	$196,240	0
11	Ceramic Tile	$22,290	$0	$22,290	0
12	Roofing and Sheet Metal	$206,712	$8,268	$198,444	4
13	Resilient Flooring	$26,085	$0	$26,085	0
14	Acoustical Tile	$32,305	$0	$32,305	0
15	Painting	$96,091	$0	$96,091	0
16	Glass and Glazing	$71,853	$0	$71,853	0
17	Terrazzo	$91,976	$0	$91,976	0
18	Miscellaneous Metals	$79,642	$0	$79,642	0
19	Finish Hardware	$55,178	$0	$55,178	0
20	Plumbing, Heating, Air Cond.	$622,064	$55,986	$566,078	9
21	Electrical	$392,160	$19,608	$372,552	5
22	Clean Glass	$3,228	$0	$3,228	0
23	Paving, Curb, and Gutter	$81,724	$0	$81,724	0
		$3,602,138	$375,276	$3,226,862	
24	Change Order No. 1	$5,240	$0	$5,240	0
	Total Contract Amount	$3,607,378	$375,276	$3,232,102	10.4%

A	Cost of Work Performed to Date	$375,276
B	Materials Stored on Site (Schedule Attached)	$67,699
C	Total Work Performed and Materials Stored	$442,975
D	Less 10% Retainage	$44,298
E	Net Work Performed and Materials Stored	$398,678
F	Less Amount of Previous Payments	$180,369
G	Balance Due This Payment	$218,309

*Adapted from Richard H. Clough, Construction Contracting, 5th ed., New York: Wiley, 1986. P. 282.

Figure 11.2   *Periodic estimate for partial payment, lump-sum contract.**

The cost shown in column 1 of Figure 11.2 for each work item consists of the direct expense of the item, plus a share of the cost of job overhead, taxes, markup, and bond. The cost of these four items, often called "job burden," may be on a strictly pro rata basis, or there may be some unbalancing. It is common practice to include a disproportionate share of the burden in those items of work that are completed early in the construction period. This procedure, called "front end loading," serves the purpose of helping to reimburse the contractor for its initial costs of moving in, setting up, and commencing operations. If the owner will accept a specific pay item for move-in costs, then such unbalancing of the cost breakdown is not necessary.

11.5 PAY REQUESTS FOR LUMP-SUM CONTRACTS

Figure 11.2 illustrates the form of a typical pay request for a lump-sum building contract. Usually prepared by the contractor, it includes all subcontracted work as well as that done by the contractor's own forces. For each work classification that it does itself, the contractor estimates the percentage completed and in place. From invoices submitted by its subcontractors, suitable percentage figures are entered for all subcontracted work. These percentages are shown in column 4 of Figure 11.2. The total value of each work classification is multiplied by its percent completion and these figures are shown in column 2. To the total of completed work is added the value of all materials stored on the site, but not yet incorporated into the work. The cost of stored materials includes that of the subcontractors and is customarily set forth in a supporting schedule. From the total of work in place and materials stored on site is subtracted the retainage. This gives the total amount of money due the contractor up to the date of the pay request. From this is subtracted the amount of progress payments already made. The resulting figure gives the net amount now payable to the contractor.

Although the pay request procedure for lump-sum contracts has been in general use for many years, it has one serious defect. As shown in Figure 11.2, the project is divided for payment purposes into major work classifications, most of which are extensive and often extend over appreciable portions of the construction period. This situation can make it difficult to estimate accurately the percentages completed of the various work categories. Actual measurement of the work quantities accomplished to date is the key to accurate percentage figures, but this can become very laborious and therefore most of the percentages are established by a visual appraisal and negotiations between the project manager and the architect–engineer or owner. Contractors want these estimates to be fair representations of the actual work achieved, but understandably, they do not want them to be too low. Hence most of their percentage estimates are apt to be on the generous side.

This circumstance continues to produce vexing problems for both the contractor and the architect–engineer or owner. If it is difficult for the contractor to estimate the completion percentages accurately, it is at least equally difficult for the architect–engineer or owner to check these reported values. This presents the architect–engineer with a difficult problem because, in the interest of its client, it must make an honest effort to see that the monthly payments made to the contractor are reasonably representative of the actual progress of the job. In addition, architect–engineers have been sued by sureties in cases where defaulting contractors had received excessive progress payments. In such cases, the surety has claimed that the architect–engineer was negligent in approving progress payments that were substantially in excess of the value of work actually accomplished. Architect–engineers are at times casual with the processing of a pay request, feeling that a delay in payment will offset the generous nature of the completion percentages provided by the contractor. Although effective, this is often at odds with contract provisions regarding payment and hardly gets to the basis of the problem.

Although retainage helps to protect against excessive owner payments, it is not inconceivable, nor probably unusual, that contractors' progress payments are occasionally more than they should be. To protect its client and itself, the architect–engineer will occasionally delay payment to the contractor or reduce the amount of payment requested. The unfortunate part about this entire matter is that the architect–engineer is acting more on hunch or intuition than on solid evidence of inflated payment figures.

11.6 USE OF TIME-CONTROL ACTIVITIES FOR PAY REQUESTS

One possible answer to the pay request dilemma on lump-sum jobs is the use of time-control network activities or groups of activities for the project cost breakdown. Although this scheme undoubtedly increases the work necessary to make the initial cost breakdown and increases the length of the individual pay request, the advantages can justify the additional effort required. If direct costs have been assigned to each activity by the contractor for cost-control purposes, the activity format for pay requests is a natural extension of the basic cost system. If activities are not used for cost control, the direct cost (labor, equipment, materials, and subcontracts) of each activity, or for an activity group, can be obtained from the estimating sheets. It is an easy matter to distribute the project burden among the activities. Perfection is not required, and a certain amount of "loading" of the first activities will probably be done anyway to help the contractor recover more quickly its substantial mobilization costs.

When the total cost of each activity is known, the routine weekly progress reports can serve as a convenient and accurate basis for the monthly pay requests. For payment purposes, all activities finished by the end of the month would be reported as 100 percent complete. The percentages complete reported for the activities in progress as of the end of the month would be the same as those already available from the latest weekly progress report (Figure 9.5).

The advantages of using activities for pay requests on lump-sum projects are several. From the contractor's point of view, although more pay items are involved, it is an easy matter to compile the total cost of the work accomplished to date. No additional field measurements or inspections are required. The time-control information already available is all that is needed. The compilation of the total value of the work completed to date is more accurate than it would be otherwise, and the cost figures for each pay item and the percentages complete are much easier to check. This can eliminate many of the most troublesome aspects of monthly pay requests. On lump-sum jobs, basing pay requests on network activities seems to have much to commend it, and this procedure is now being used by some public agencies.

11.7 PAY REQUESTS FOR COST-PLUS CONTRACTS

Negotiated contracts of the cost-plus variety provide numerous methods for making payments to contractors. In many cases, contractors furnish their own capital, receiving periodic reimbursement from owners for costs incurred. Other contracts provide that owners will advance the contractors money to meet payrolls and to pay other expenses associated with the work. One scheme is where the contractor prepares estimates of expenses for the coming month and receives the money in advance. Then, at month's end, the contractor prepares an accounting of its actual expenses. Any difference between the estimated expenses and actual expenses is adjusted with the next monthly estimate. Other contracts provide for zero-balance or constant-balance bank accounts where checks are written by the contractor and funds are furnished by the owner. In any event, the contractor must make periodic accountings to the owner for the cost of the work, either to receive direct payment from the owner or to obtain further advances of funds.

The matter of periodic payments to the contractor under a cost-plus type of contract is not based on quantities of work done, but on expenses incurred by the contractor in prosecuting the work. Consequently, such pay requests consist primarily of the submission of original cost records. Invoices, payrolls, vouchers, and receipts are submitted in substantiation of the contractor's payment requests. In addition to cost records of direct expense, the periodic pay requests customarily include a pro rata share of the negotiated fee. If the contractor receives advances for construction costs, the owner is customarily credited with all cash discounts.

Because of the sensitive nature of cost reimbursement, it is common practice to maintain a separate set of accounting records for each cost-plus project. When the size of the project is substantial enough to justify it, a field office is established where all matters pertaining to payroll, purchasing, disbursements, and record keeping for the project are performed. Project financial records are either routed through the owner's representative or are available for inspection at any time. This procedure does much to eliminate misunderstandings and facilitates the final audit.

11.8 PAYMENTS TO SUBCONTRACTORS

When the general contractor receives monthly invoices from its subcontractors, it has its own problems of verifying the requested amounts. In many instances, the general contractor does not require nor receive cost breakdowns from its subcontractors, and the project manager must expend considerable time and effort in checking subcontractor invoices. Even then, unless he happens to be experienced in each construction specialty, the project manager usually has no real basis for accurately evaluating the progress of a subcontractor's work on a project.

Many prime contractors require their subcontractors to submit appropriate cost breakdowns for payment purposes. These cost schedules are used by the contractors to evaluate the pay requests submitted by their subcontractors. One form of price breakdown that can be used is to have each subcontractor place a price tag on every network activity with which it is involved. It is not difficult for the project manager to check the reasonableness of the reported amounts for each activity. Once the contractor and the subcontractor have agreed that prescribed sums are due on the completion of designated activities, the analysis and verification of monthly invoices from subcontractors are readily performed.

General contractors are anxious to prepare their requests for payment and transmit them to the owners as soon after the first of the month as possible. With a breakdown of subcontractor prices for each activity, the project managers can determine the amount due each subcontractor from their own evaluation of activity progress. This evaluation can be substantiated with the subcontractor and placed on the prime contractor's request for payment. This can reduce the time necessary to prepare the pay estimate and send it on to the owner. Subcontractors also benefit because earlier payment to the general contractor means earlier payment to them.

11.9 SCHEDULE OF PAYMENTS BY OWNER — UNIT-PRICE CONTRACT

A common contract provision requires the contractor to provide the owner before the start of construction with an estimated schedule of monthly payments that will become due during the construction period. This information is needed by the owner so that it can have cash available to make the necessary periodic payments to the contractor. Because the owner must sometimes sell bonds or other forms of securities to obtain funds with which to pay the contractor, it is important that the anticipated payment schedule be as accurate a forecast as the contractor can make it.

When a unit-price contract is involved, the payment schedule depends entirely on the quantity of each bid item that will have been completed each month during the construction period. The highway bridge was bid on the basis of unit prices and Figure 3.9 is a schedule of the bid items and their respective quantities and bid amounts. Once the job schedule has been established, it is an easy matter to make a compilation of total bid-item quantities that are scheduled to be completed as of the end of each month. Using Figure 5.14, the schedule of progress payments shown in Figure 11.3 is made quickly and directly.

In Figure 11.3, it is to be noted that the total value of work accomplished as of the end of each month is compiled on the basis of bid-item quantities done and their bid unit prices. Any variation between the actual quantities and the estimated quantities will make the schedule of payments inaccurate and will result in a future modification of the payment schedule. It has been assumed on the highway bridge that a retainage of 10 percent will be withheld by the owner until contract completion. In accordance with this, the total amount due the contractor as of the end of each month is multiplied by a factor of 0.90. Because the highway bridge is to be completed in September, the

Item No.	Bid Item	Estimated Quantity	Bid Unit Price	June 30 Estimated Quantity Complete	June 30 Total Value to Date	July 31 Estimated Quantity Complete	July 31 Total Value to Date	August 31 Estimated Quantity Complete	August 31 Total Value to Date	September 30 Estimated Quantity Complete	September 30 Total Value to Date
1	Excavation, unclassified	1,667	$2.10	418	$878	1667	$3,501	1,667	$3,501	1667	$3,501
2	Excavation, structural	120	$29.53			120	$252	120	$3,544	120	$3,544
3	Backfill, compacted	340	$11.40					340	$3,876	340	$3,876
4	Piling, steel	2,240	$34.02			2240	$4,704	2,240	$76,205	2240	$76,205
5	Concrete, footings	120	$113.62			120	$252	120	$13,634	120	$13,634
6	Concrete, abutments	280	$240.65			140	$294	280	$67,382	280	$67,382
7	Concrete, deck slab, 10 in.	200	$99.03					200	$19,806	200	$19,806
8	Steel, reinforcing	90,000	$0.65			57150	$120,015	90,000	$58,500	90000	$58,500
9	Steel, structural	65,500	$0.69					65,500	$45,195	65500	$45,195
10	Bearing plates	3,200	$1.66					3,200	$5,312	3200	$5,312
11	Guardrail	120	$65.95					40	$2,638	120	$7,914
12	Paint	job	$8,394.55					20%	$1,679	100%	$8,395
				Totals	$878		$129,018		$301,271		$313,263

		Monthly progress payment	Total payment on contract
June 30	$878 x 90% =	$790	$790
July 31	$129,018 x 90% - $790 =	$115,326	$116,116
August 31	$301,271 x 90% - $116,116 =	$155,028	$271,144
September 30	$313,263 x 90% - $271,144 =	$10,792	$281,937
Final Payment		$31,326	$313,263

Figure 11.3 Highway bridge, schedule of progress payments.

payment for that month has been shown to be the final payment, including accumulated retainage. In fact, the monthly payment amounts obtained in Figure 11.3 will not usually be paid at the end of the month. Rather, construction contracts frequently provide that the owner shall make payment to the contractor 10 days after a suitable pay request has been submitted. Contractors normally submit such pay requests at the end of each month so payment by the owner is not made until about the tenth of the month following.

The monthly payment schedule presented in Figure 11.3 was prepared with all activities starting as early as possible with the exception of activity 40, "Move in," and the following three activities 80, 90, and 100. As discussed earlier in Section 8.5 and Section 9.4, the starts of these four activities are being purposely delayed by nine working days. It will be recalled that this move will not delay the project in any way and will eliminate a substantial period during which the project would otherwise be at a complete standstill awaiting delivery of steel pilings and reinforcing steel. It has been amply demonstrated in this text that, although early start schedules serve as optimistic goals, there are many practical reasons why such schedules must often be modified. Deviations from the early start schedule are almost always in the nature of delays and later starts. As a result of this, the monthly payment schedule obtained in Figure 11.3 may well indicate the early monthly payments to be too high and the later payments to be too low. If monthly payment schedules based on an early start schedule are submitted to the owner, the first few pay requests submitted by the contractor may be smaller than those forecasted because of inconsequential schedule slippages. This is apt to lead the owner and architect–engineer to conclude that the work is falling behind schedule, even when things are going along very well.

If schedule adjustments have to be made to effect project shortening or leveling of resources, the monthly payment schedule can be made to reflect these. Otherwise, making allowances for the inevitable schedule slippages is very much a matter of intuition and judgment. One approach to this matter would be to compile a payment schedule on the basis of a late start schedule. That is to say, assume that every activity will start as late as possible. This would, of course, produce a payment schedule with too low initial payments and too high later ones. Presumably, the actual payment schedule will be somewhere between the early start and the late start payment schedules.

11.10 SCHEDULE OF PAYMENTS BY OWNER — LUMP-SUM CONTRACT

The payment schedule for a lump-sum project is often computed using the traditional bar chart, one that is not network based. An approximate cost is established for each bar on the chart, and this cost is distributed uniformly over the length of the bar. From this, a cumulative total project cost curve is computed for the contract period. The problem with this procedure is the inaccurate way the costs of the project segments are distributed over the contract period. Although the total cost of each major job segment can be established with reasonable accuracy, the time rates at which these expenses are incurred can involve such variation that the payment schedule derived therefrom may be seriously in error. To illustrate, bar charts often have a category called "Electrical" or "Mechanical" whose bar extends from project start to finish. The total value of this work, which is usually subcontracted, is distributed uniformly over its duration. This is seldom a realistic time allocation of expense.

A much more dependable procedure is to compute cumulative job cost values using the project schedule and cost of time-control activities as a basis. Closely associated activities can be combined for simplicity. The total cost (direct cost, plus a pro rata share of job overhead, tax, markup, and bond) of each activity is required. This is the same activity cost that would be needed for monthly payment purposes in Section 11.6. Dividing the total cost of each activity by its estimated duration gives the cost of the activity per working day. Because the duration of each activity is short compared with a monthly pay period, the error induced by the uniform distribution of cost is negligible. Using the activity cost figures and the project schedule, the cumulative project expense can be compiled. One way this can be accomplished is to use a worksheet with a horizontal time scale in terms of working days. The activities can be listed vertically with the daily cost of each activity entered for the appropriate work days. The costs of activities involving the procurement of job materials are charged lump sum on the last day of their duration. Adding the costs cumulatively from left to right produces the daily total project cost to date.

Although the process just described produces a cumulative total project cost up through each working day, only those costs as of the end of each month are usually of real concern. Therefore, an alternative procedure can be used that is more direct. As of the end of each month, simply add up the values of all activities scheduled to be completed on or before the end of that month. For activities in process as of that time, include an appropriate percentage of their costs. This is a rapid, accurate way to determine the total value of the work that will have been accomplished as of the end of any specified month.

11.11   FINAL PAYMENT

The steps leading up to acceptance of the project and final payment by the owner vary with the type of contract and the nature of the work involved. A typical procedure is where the contractor, having achieved substantial completion, requests a preliminary inspection. The owner or its authorized representative, in company with general contractor and subcontractor personnel, inspect the work. A "punch list" is made up describing all the observed deficiencies. After the work has been finalized and all deficiencies remedied, a final inspection is held. Following this, the owner makes formal written acceptance of the project and the contractor presents its application for final payment. Under a lump-sum form of contract, the final payment is the final contract price less the total of all previous payment installments made. With a unit-price contract, the final total quantities of all payment items are measured and the final contract amount is determined. Final payment is again equal to the contract price less the sum of all progress payments previously made. In all cases, final payment by the owner includes all retainage that has been withheld.

Construction contracts often require that the contractor's request for final payment be accompanied by a number of different documents. For example, releases or waivers of lien executed by the general contractor, all subcontractors, and material suppliers are common requirements on privately financed jobs. Other contracts call for an affidavit certifying that all payrolls, bills for materials, payment to subcontractors, and other indebtedness connected with the work have been paid or otherwise satisfied. Construction contracts frequently require the contractor to provide the owner with as-built drawings, various forms of written warranties, maintenance bonds, and literature pertaining to the operation and maintenance of job machinery. Consent of surety to final payment is an almost universal prerequisite. The project manager is responsible for getting the work accepted by the owner and conforming with contractual provisions pertaining to final payment.

11.12 CASH FLOW

Construction projects can make substantial demands on a contractor's cash. Initially incurred are the usual start-up costs of moving in workers and equipment; erecting a field office, storage sheds, fences, and barricades; job layout; and installation of temporary electrical, water, telephone, sanitary, and other services. The premiums for performance and payment bonds as well as for certain types of permits and project insurance must be paid at the inception of field operations. There is seldom a pay item pertaining specifically to these start-up costs, and the contractor recovers these expenses only as the work progresses.

The contractor's investment in the job is increased even further after the work gets underway. It must meet its payroll costs on a weekly basis and will want to take advantage of cash discounts when paying material bills if it possibly can. The contractor does receive progress payments at monthly intervals from the owner. However, these payments are not normally due until sometime during the month following their submittal and their amounts are reduced by retainage.

As a consequence of these circumstances, the contractor's expense on a project will typically exceed its monthly progress payment income over an appreciable part of the construction period. The cash deficit on the project must be made up from the contractor's working capital, or money must be borrowed to provide the necessary operating funds. "Cash flow" refers to a contractor's income and outgo of cash. The net cash flow is the difference between disbursements and income at any point in time. A negative net cash flow means disbursements are exceeding income, a normal situation on even a highly profitable project during the greater part of its duration. A determination of the future rates of cash disbursements and cash income together with their combined effect on the project cash balance is called a "cash flow forecast."

The comptroller or financial vice-president of a construction company is concerned with the combined effects of the cash flow forecasts of all of the company's projects. Since the cash flow forecast for the company is simply the sum of the forecasts for the individual projects, it is the responsibility of the project manager to determine the cash flow of his project and to make regular revisions to it as the job progresses.

Cash flow forecasting is equally important for both large and small construction companies. Large firms with many concurrent projects try to use the positive cash flows of one project to handle the negative cash flows of another. Where cash demands exceed the normal working capital, arrangements for short-term loans are made and repayment schedules established. Where cash income exceeds the demand, short-term investments are made, with a liquidation schedule to fit future demands. Cash forecasting can be done with no more than approximate accuracy, but it is a useful device for controlling a company's cash position. The corporate profit of a large company can be greatly enhanced by proper cash management.

Small construction companies as well as growing firms have a special cash flow situation. Working capital in these companies is almost always in short supply. There are continual requirements for additional equipment and tools to handle the growing size and number of projects. Lending institutions are hesitant to make large loans to small companies. This is especially true when the funds are needed for working capital. As a result, the growth of small firms is often limited by their cash flow position. The ability to forecast accurately cash flow needs and to manage cash as a resource can increase a firm's growth rate and its annual turnover of projects.

As workers work and materials are purchased on a construction project, their costs accumulate on the contractor's books. These costs are referred to as accrued costs. When payment is made for the labor and materials, these are cash disbursements. Accrued costs and cash disbursements are exactly equal in value, but their timing is different. Generally, disbursements follow accrued costs.

As workers work and materials are put into place, the value of a construction project increases. This increase in the value of the construction represents an accrued income to the contractor. Periodically, the contractor and the architect–engineer evaluate the value of the construction, and the contractor prepares a progress payment request. When the owner sends a check to the contractor, the contractor's accrued income becomes a cash receipt. Accrued income and cash receipts are equal in value, but as in the case of accrued costs and cash disbursements, their timing is different. In this case, the cash receipts usually follow the accrued income.

It is the intent of the contractor to accrue income faster than it accrues costs. In the end, this difference represents the contractor's profit. The problem is that the cost of the work put into a project is not directly related to the resulting income. Additionally, the time delays between the accrued costs and the cash disbursement are not the same as the time delays between the accrued income and the cash receipts. To forecast the amount of money the contractor must invest in the project, it is necessary to estimate the amount and the timing of the cash disbursements and the cash receipts.

11.13 CASH DISBURSEMENT FORECASTS

Cash disbursements by contractors on construction projects can be divided into three classifications. First, there are the "up-front" costs or initial expenses necessary to start the project. These include various payments that must be made before construction starts such as the costs of bonds, permits, insurance, and expenses of a similar kind. The second group of disbursements involves the payment of direct job expenses. These include costs associated with payrolls, materials, construction equipment, and subcontractor payments. The third classification relates to payments for field overhead expense and tax.

The original cost estimate and project schedule provide the basic cost and time information needed to make a cash disbursement forecast. The monthly cash disbursements for the highway bridge are calculated in Figure 11.4. The only up-front cost is $3,752 for the performance and payment bonds. The direct costs for each project activity are listed. The costs of activity 40 ($8,773) and of activity 390 ($3,940) are overhead expenses and their sum of $12,713 is deducted from the total job overhead of $41,223. (See Figure 3.7). As shown in Figure 11.4, the total of this net project overhead, small tools ($2,809) and tax ($7,825) amounts to $39,144. This sum is spread uniformly over the project duration of 70 working days at the rate of $559 per day.

Reference is made to Figure 11.4 and the computation of the contractor's disbursements that will be made on the highway bridge for the month of June. The initial expense for the contract bonds is shown as $3,752. Job overhead for the first 13 working days is $7,267. By June 30 (project day 13), activities 40 and 70 are scheduled for completion and activities 8, 90, and 100 will have been in process for 1 day. The direct cost of these activities is $20,410. The sum of these different expenses is $31,429, which is the predicted total contractor outlay on the highway bridge as of the end of June. Similar computations are presented for the months of July, August, and September. These cumulative project costs are plotted and connected by a smooth "S" curve in Figure 11.5. Because the contractor pays its bills throughout each month, the smooth expense curve shown in Figure 11.5 is considered to be an acceptable representation of project cash disbursements.

This procedure for calculating the time rate of disbursement on a construction job involves many approximations. Greater accuracy could be obtained by separating the direct job expenses into labor, materials, equipment, and subcontractor categories. For each of these cost components, different delays would apply between when the expense is incurred and when it is actually paid. Such a procedure is extremely complicated, however. The assumption of a zero time delay for all direct costs, as was done in Figures 11.4 and 11.5, is conservative and suitably accurate for most purposes.

11.14 CASH INCOME FORECASTS

Cash income to the contractor consists of progress payments from the owner. Estimated amounts of the monthly payments for the highway bridge, based on an early start schedule, have already been obtained in Figure 11.3 and are shown in Figure 11.5. However, as has already been pointed out, these payments are not received by the contractor until about the tenth of the following month. Consequently, when the values from Figure 11.3 are plotted in Figure 11.5, they must be displaced to the right by seven working days. Notice that the cash income curve forms a sawtooth with income occurring only once each month.

Activity No.	Description	Duration	Direct Cost		Cost	Cumulative Cost
30	Order & deliver piles	15	$30,490	June 30	$31,429	$31,429
40	Move in	3	$8,773	Bond	$3,752	
60	**Fabricate & deliver abutment & deck rebar**	**15**	$15,055	13 days indirect @ $559 per day	$7,267	
70	Fabricate & deliver footing rebar	7	$5,645	Activities 40, 70	$14,418	
80	Prefabricate abutment forms	3	$4,552	Activity 80 1 day	$1,517	
90	Excavate abutment #1	3	$2,845	Activity 90 1 day	$948	
100	Mobilize pile driving rig	2	$7,054	Activity 100 1 day	$3,527	
110	Drive piles abutment #1	3	$5,880	July 31	$121,887	$153,316
120	Excavate abutment #2	2	$2,036	21 days indirect @ $559 per day	$11,739	
130	Forms & rebar footing #1	2	$4,037	Activities 30, 60, 110, 120, 130, 140, 150,		
140	Drive piles abutment #2	3	$5,879	160, 170, 180, 190, 200, 210, 220, 230	$101,689	
150	Pour footing #1	1	$3,830	Activity 80 2 days	$3,034	
160	Demobilize pile driving rig	1	$3,527	Activity 90 2 days	$1,896	
170	Strip footing #1	1	$92	Activity 100 1 day	$3,527	
180	**Forms & rebar abutment #1**	**4**	$9,541			
190	Forms & rebar footing #2	2	$2,857			
200	**Pour abutment #1**	**2**	$12,180	August 31	$100,842	$254,158
210	Pour footing #2	1	$3,830	22 days indirect @ $559 per day	$12,298	
220	**Strip & cure abutment #1**	**3**	$2,363	Activities 240, 250, 260, 270, 280, 290,		
230	Strip footing #2	1	$92	300, 310, 320, 330, 340, 350, 380	$86,715	
240	**Forms & rebar abutment #2**	**4**	$9,541	Activity 360 1 day	$1,829	
250	**Pour abutment #2**	**2**	$12,179	Activity 370 1 day	$1,164	
260	Fabricate & deliver girders	25	$29,192	September 30	$20,080	$274,238
270	Rub concrete abutment #1	3	$1,287	14 days indirect @ $559 per day	$7,826	
280	Backfill abutment #1	3	$1,344	Activity 360 2 days	$3,658	
290	**Strip & cure abutment #2**	**3**	$2,362	Activity 370 4 days	$4,656	
300	Rub concrete abutment #2	3	$1,286	Activity 390	$3,940	
310	Backfill abutment #2	3	$1,344			
320	**Set girders**	**2**	$5,739			
330	**Deck forms & rebar**	**4**	$9,889			
340	**Pour & cure deck**	**3**	$7,536	Job overhead	$41,223	
350	**Strip deck**	**3**	$1,789	Small tools	$2,809	
360	Guardrails	3	$5,487	Less Move in and Cleanup	($12,713)	
370	**Paint**	**5**	$5,820	Tax	$7,825	
380	Saw joints	1	$234		$39,144	
390	**Cleanup**	**3**	$3,940		÷70 days = $559	

Figure 11.4   Highway bridge, contractor's expense.

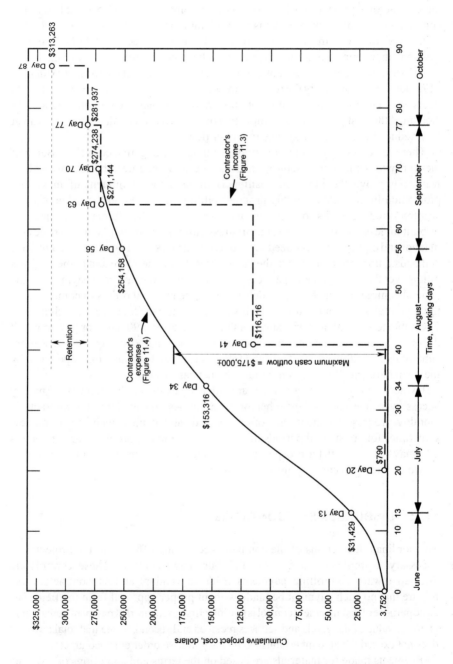

Figure 11.5 *Highway bridge, contractor's expense and income.*

The curves in Figure 11.5 are typical of most construction projects. At any point in time, the vertical distance between the two curves represents the amount by which the contractor's project expense exceeds its receipts. Figure 11.5 reveals that the contractor on the highway bridge has a considerable amount of money invested in the owner's job for the entire construction period. It is not until receipt of the last progress payment on working day 77 that the cash flow becomes positive. At one point during the construction period the contractor's cash deficit amounts to approximately $175,000. This is precisely the reason why some unbalancing, either in unit-price bid items or in the cost breakdown on lump-sum jobs, is such a common practice. By "loading" the first pay items accomplished on a given project, the contractor can at least reduce the amount of its negative cash flow.

There are other ways of minimizing the amount of negative cash flow. Some of these are within the control of the project manager and some are not. Probably the best way to improve the job cash position is through the attainment of maximum production in the field. Monthly progress billings are based on the value of work accomplished, not on its cost (except on cost-plus jobs). Therefore, if the project can be built for less than the estimate, the progress billings remain as planned and the cost of labor and equipment is reduced. Subcontracting work has a positive effect on cash demands. Subcontractors add to the value of work and therefore add to the progress billing, yet they are typically paid only after receipt of the progress payment by the general contractor. The timing of delivery of large material orders to coincide with the submittal of the contractor's monthly pay estimate can reduce cash requirements. Favorable terms with material suppliers involving 60 and 90 days for payment will decrease cash requirements, but may have an adverse effect on profit. Prompt submittal of monthly pay requests and follow-up action to ensure timely payment by the owner are excellent ways to reduce the net cash flow.

Owners and architect–engineers are often unaware of the amount of money necessary to meet payrolls and other job expenses. As a result of this, the contractor's monthly progress payments are often delayed, and it must wait for its money, sometimes for substantial periods of time. This is a major problem for contractors who may be faced with the prospect of borrowing large amounts of short-term money at two or three percentage points above prime rate.

11.15 DISBURSEMENT CONTROLS

To coordinate the actions of the company accounting office with the project, it is necessary to implement a system of disbursement controls. These controls are directed toward controlling payments made to vendors and subcontractors and require that no such payment be made without proper approval from the field. The basic purpose of disbursement control is twofold: first, to ensure that payment is made only up to the value of the goods and services received and second, to see that total payment does not exceed the amount established by the purchase order or subcontract.

Payments made for materials are based on the terms and conditions of covering purchase orders. Copies of all job purchase orders are provided to the project manager for his use and information. Purchase order disbursement by the accounting office is conditioned on the receipt of a signed delivery ticket or receiving report from the job site. Suitable internal controls are established to ensure that total payments do not exceed the purchase order amount. Any change in purchase order amount, terms, or conditions is in the form of a formal written modification with copies sent to the job site.

Disbursements to subcontractors follow a similar pattern. Because there are no delivery tickets or receiving reports for subcontractors, all subcontractor invoices are routed for approval through the project manager who has copies of all the subcontracts. The project manager compares the invoice with his progress determination and approves the invoice or makes appropriate changes. The subcontractor is not actually paid until the owner has made payment to the general contractor. General contractors normally withhold the same percentage from their subcontractors that owners retain from them. If the subcontractor bills for materials stored on site, a common requirement is that copies of invoices be submitted to substantiate the amounts billed. Any change to a subcontract is accomplished by a formal change order.

11.16 PROJECT CHANGES

Changes in the work or deviations from the anticipated job site conditions can stem from a variety of causative factors. The owner or architect–engineer may decide to add additional work or change certain contract requirements. The contractor may suggest construction changes in accordance with the contract's value-engineering clause. The owner, architect–engineer, or another prime contractor may cause delay in the prosecution of the field work. Job-site conditions may be encountered that are appreciably different from those originally contemplated. Changes of this sort can result in work extra to the contract, extensions of contract time, and claims for additional costs.

The usual construction contract is explicit about how changes are to be handled and the extent of the owner's liability. The project manager has an important responsibility in evaluating the effects of project changes and taking all the steps necessary to protect the contractor's interests. This involves the negotiation of contract change orders, keeping detailed job records, analyzing the time and cost effects of project changes, documentation of extra costs, and timely notification to the owner of all job delays, extra costs, and claims under the contract. In particular, the job manager must be especially careful to proceed in full conformance with the applicable contract provisions.

The time and cost consequences of project changes are often difficult to document, especially in a form meaningful to the owner or architect–engineer or perhaps that can serve as evidence in arbitration or litigation. The project network is a powerful tool for analyzing the time effects of a project change. This matter has been previously discussed in Section 9.21. In a situation where a substantial change in the work occurs, it is advisable to update the network so that it is completely current, but without reflecting the proposed change. Network calculations and a current cost forecast report are used to establish the time and cost status of the project before the change occurs. The project network is then modified to reflect the change in the work.

A new set of network calculations is made and the cost implications are figured into a revised cost forecast report. Any differences in project time and cost are the result of the proposed change.

Care must be taken in interpreting the results of this procedure. A change in the work can have far-reaching implications. As an example, a change in the foundation of a building project can seriously delay all of the work that follows. This may mean that the project cannot be closed in before winter weather, further delaying other trades. Major items of material and project equipment may already be ordered and must be received and stored prior to their need. Job delays can cause work to be rescheduled from periods of favorable to unfavorable weather. On the highway bridge, a delay in the start of the project or trouble with an abutment foundation could push the work into the rainy season where the normally dry river bed becomes a swift stream. Delays in some cases can improve a bad timing situation, but generally the project plan has been designed to take advantage of weather and other conditions as much as possible. Therefore, delays usually tend to disrupt a project plan, making replanning more difficult and resulting in higher costs to the contractor.

It is always advisable to analyze the time and cost implications of a project change at the time that it occurs. For one reason, any revision made to the construction contract to incorporate the change must be based on an accurate assessment of the total effect of the change. Another reason for making a prompt study is that job changes can involve substantial time and cost and occasionally lead to disputes between owner and contractor that are not settled until long after the project is finished. An analysis made long after the fact to demonstrate the effect of a job change produces data of questionable value and authenticity at best.

11.17   CONTRACT CHANGE ORDERS

Alterations to the contract that involve modifications to the time or price of the project are consummated by formal change orders. These changes may alter the contract by additions, deletions, or modifications to the work and can be initiated by the owner, architect–engineer, or contractor. The dollar amount of a change is negotiated and, depending on the contract terms, can be expressed as a lump sum, unit prices, or cost plus a fee. A number of changes are sometimes incorporated into one change order, but each change needs to be documented at its inception, estimated for cost and time consequences before work commences, and written approval to proceed received from the owner or architect–engineer. Strict adherence to this change policy greatly reduces misunderstandings.

An important part of a change order is any extension of contract time that may be required as a consequence of the change. In the absence of a project network, the contractor has no real basis for determining the additional contract time required and will usually either request no extra time or an extension equal to the full time required to accomplish the extra work itself. As was discussed in Section 11.16, the influence of the change on total project duration can be clearly demonstrated by performing a forward pass on the current network and a forward pass with the change incorporated into the network. Often, the change affects only a noncritical path and no additional time is justified. Even so, the contractor's scheduling leeway has been decreased with possible cost implications. If a longer critical path results, the net additional time actually required can be easily determined. Such a project network analysis can be very effective in substantiating a claim for additional contract time.

Part of the change order dollar amount is an allowance for job overhead. Many contracts provide that the cost of extra work shall include an amount for overhead expressed as a fixed percentage (10 to 15 percent is common) of the direct cost involved. Basically, this is an attempt to reimburse the contractor for its office expense of estimating the cost of the change and processing the paperwork rather than an allowance for additional field overhead expense. This is a satisfactory arrangement when no increase in project time is involved, but can be most unsatisfactory when the duration is extended. The job overhead on the highway bridge is $559 per day, and on larger projects it can be a major expense. Network analysis can provide the number of days of additional project overhead that should be charged when such a change is being negotiated. Where a deductive change order is concerned, only direct costs will be involved unless the project duration is actually decreased.

11.18 CLAIMS

During the construction period, disputes sometimes arise between the owner and the contractor concerning claims by the contractor for extensions of time or payment of extra costs. If such claims cannot be amicably settled during the construction period, they must either be dropped by the contractor or be decided by arbitration, appeal boards, or the courts.

Claims of this type can stem from a variety of conditions and often involve substantial sums of money. Job delays caused by the owner or another of its prime contractors result in many claims against the owner for "impact costs" or "consequential damages." Impact costs are additional expenses incurred by the contractor as a consequence of a delay to the project. Claims for extensions of time and extra costs often result from failure of the owner to furnish sites, make timely decisions, or provide owner-furnished materials. Errors or oversights on the part of the architect–engineer account for numerous contractor claims.

Another category of claims can result from the contractor's dealings with its subcontractors. The coordination and timing of the work of subcontractors is critical to the overall job schedule. A delay in the work of one subcontractor can have a domino effect on the work of everyone that follows. It was mentioned earlier that subcontractors should have a definite voice in the preparation of the original schedule. This tends to make the schedule realistic from the standpoint of all the parties who will be responsible for its execution and at the same time will show individual subcontractors how their work fits into the overall construction program. The project network, when kept updated, shows when all subcontractors start and complete their work and what effect any delay on their part will have on the work yet to be done. The updated diagram provides an excellent means of allocating responsibility between delaying parties and assigning the resulting financial responsibility.

Disputes and claims are commonplace in the construction industry. To assist the contractor in pursuing its own claims and in defending itself against those made against it, the byword is **documentation**. If the contractor's position is to prevail in such matters, full and complete documentation of all pertinent facts and information is an absolute necessity. The as-built diagram discussed in Section 9.22 is one important element of a contractor's project documentation process. The standard dictum of "put everything in writing" is very important. Letters, memoranda, drawings, notes, diaries, photographs, and clippings can be useful. Basic to job documentation is the daily job log.

11.19   DAILY JOB LOG

A job log is a historical record of the daily events that take place on the job site. The information to be included is a matter of personal judgment, but should include everything relevant to the work and its performance. The date, weather conditions, numbers of workers, and amounts of equipment should always be noted. It is advisable to indicate the numbers of workers by craft and to list the equipment items by type. A general discussion of the daily progress, including a description of the activities completed and started and an assessment of the work accomplished on activities in progress, is important. Where possible and appropriate, quantities of work put into place can be included.

The diary should list the subcontractors who worked on the site together with the workers and equipment provided. Note should be made of the performance of subcontractors and how well they are conforming to the project time schedule. Material deliveries received must be noted together with any shortages or damage incurred. It is especially important to note when material delivery dates are not met and to record the effect of such delays on job progress and costs.

The diary should include the names of visitors to the site and facts pertinent thereto. Visits by owner representatives, the architect–engineer, safety inspectors, union representatives, and people from utilities and governmental agencies should be documented and described. Meetings of various groups at the job site should be recorded, including the names of people in attendance, problems discussed, and conclusions reached.

Complete diary information is occasionally necessary for extra work and is always needed for any work that might involve a claim. The daily diary should always include a description of job problems and what steps are being taken to correct them. The job log is an especially important document where disputes result in arbitration or litigation. To be accepted by the courts as evidence, the job diary must meet several criteria. The entries in the log must be original entries made on the dates shown. The entries must have been made in the regular course of business and must constitute a regular company record. The entries must be made contemporaneously with the events being recorded and must be based on the personal knowledge of the person making them. It is also preferable that the log not be kept looseleaf where sheets can be added or removed, but be maintained in a bound booklet or journal. Where the above criteria have been met, the courts have generally ruled that the diary itself can be used as evidence, even if the author is not available to testify.

Highway Bridge Bid-Item Summary Sheets

SUMMARY SHEET

Job: Highway Bridge   Bid Item No. 1: Excavation, unclassified   Estimator: GAS

Cost Code	Work Type	Quantity	Unit	Calculations	Labor Cost Direct	Labor Cost Indirect	Equipment Cost	Material Cost	Total Cost
02220.10	Excavation, unclassified	1,667	cy	**Labor:** 1 Foreman @ $17.20 = $17.20 1 Operator @ $14.51 = $14.51 2 Laborers @ $11.60 = $23.20 Crew hourly rate = $54.91 Production rate = 70 cy per hour 1,667 ÷ 70 x $54.91 = $1,308 **Equipment:** Tractor dozer @ $26.00 per hour 1,667 ÷ 70 x $26.00 = $619	$1,308	$497	$619		$1,805
				Total this account	$1,308	$497	$619	$0	$2,424
					$1,308	**$497**	**$619**		
					$1,805				
Total Bid Item No. 1							**$619**	**$0**	**$2,424**

SUMMARY SHEET

Job: Highway Bridge Bid Item No. 2: Excavation, structural Estimator: GAS

Cost Code	Work Type	Quantity	Unit	Calculations	Labor Cost Direct	Labor Cost Indirect	Equipment Cost	Material Cost	Total Cost
02222.10	Excavation, structural	120	cy	**Labor:** 1 Foreman @ $17.20 = $17.20 1 Operator @ $14.51 = $14.51 6 Laborers @ $11.60 = $69.60 Crew hourly rate = $101.31 Production rate = 8 cy per hour 120 ÷ 8 × $101.31 = $1,520 **Equipment:** 1 cy backhoe @ $26.00 per hour 120 ÷ 8 × $26.00 = $390	$1,520	$547	$390		$2,067 $390
				Total this account	$1,520	$547	$390	$0	$2,457
Total Bid Item No. 2					**$1,520**	**$547** **$2,067**	**$390**	**$0**	**$2,457**

SUMMARY SHEET

Job: Highway Bridge Bid Item No. 3: Backfill, compacted Estimator: GAS

Cost Code	Work Type	Quantity	Unit	Calculations	Labor Cost Direct	Labor Cost Indirect	Equipment Cost	Material Cost	Total Cost
02226.10	Backfill, compacted	340	cy	**Labor:** Labor unit cost = $4.80 per cy 340 × $4.80 = $1,632 **Equipment:** Equipment unit cost = $1.38 per cy 340 × $1.38 = $469	$1,632	$587	$469		$2,219 $469
				Total this account	$1,632	$587	$469	$0	$2,688
Total Bid Item No. 3					**$1,632**	**$587** **$2,219**	**$469**	**$0**	**$2,688**

SUMMARY SHEET

Job: Highway Bridge   Bid Item No. 4: Piling, steel

Cost Code	Work Type	Quantity	Unit	Calculations
02350.00	Piledriving rig, mobilize and demobilize	job	ls	**Labor:** 1 Foreman @ \$17.20 = \$17.20 2 Operator @ \$15.95 = \$31.90 1 Oiler @ \$12.08 = \$12.08 2 Piledrivermen @ \$17.05 = \$34.10 2 Laborers @ \$11.60 = \$23.20 2 Truck drivers @ \$13.80 = \$27.60 Crew hourly rate = \$146.08 three 8-hour days required 24 x \$146.08 = \$3,506 **Equipment:** 50 ton crane @ \$108.00 25 ton crane @ \$ 76.50 Lowboy @ \$ 30.00 Flat bed Truck @ \$ 15.00 Equipment hourly rate = \$ 229.50 24 x \$229.50 = \$5,508 **Material:** General supplies Total this account
02361.10	Piling, steel, driving	2,240	lf	**Labor:** 1 Foreman @ \$17.20 = \$17.20 1 Operator @ \$15.95 = \$15.95 1 Oiler @ \$12.08 = \$12.08 3 Piledrivermen @ \$17.05 = \$51.15 2 Laborers @ \$11.60 = \$23.20 1 Carpenter @ \$16.20 = \$16.20 Crew hourly rate = \$135.78 Production rate = 70 ft per hour 2,240 ÷ 70 x \$135.78 = \$4,345 **Equipment:** 50 ton crane @ \$108.00 600 cfm compressor @ \$ 12.38 Pile hammer & leads @ \$ 57.00 Equipment hourly rate = \$ 177.38 2,240 ÷ 70 x \$177.38 = \$5,676 **Material:** Steel Piling 2,240 x \$13.55 = \$30,352 Pile cushions 2,240 ÷ 120 x \$7.40 = \$138 Total this account

Total Bid Item No. 4

Labor Cost		Equipment	Material	Total
Direct	Indirect	Cost	Cost	Cost
				Estimator: GAS
$3,506	$1,367			$4,873
		$5,508		$5,508
			$200	$200
$3,506	$1,367	$5,508	$200	$10,581
$4,345	$1,738			$6,083
		$5,676		$5,676
			$30,352	$30,352
			$138	$138
$4,345	$1,738	$5,676	$30,490	$42,249
$7,851	**$3,105**			
	$10,956	**$11,184**	**$30,690**	**$52,830**

SUMMARY SHEET

Job: Highway Bridge   Bid Item No. 5: Concrete, footings

Cost Code	Work Type	Quantity	Unit	Calculations
03150.10	Footing forms, fabricate	360	sf	**Labor:** Labor unit cost $1.85 per sf 360 x $1.85 = $666.00 **Material:** Plyform: 2 uses, 50% salvage 360 x $0.47 x 0.5 = $85.00 Lumber: 1.33 fbm per sq. ft., 2 uses, 50% salvage 360 x 1.33 x $0.45 x .5 = 107.73
				Total this account
03157.10	Footing forms, place & strip	720	sf	**Labor:** Labor unit cost = $0.48 per sf 720 x $0.48 = $345.60 (Approx. 70% for placing, 30% for stripping) **Material:** Nails, hardware, coatings 720 x $0.18 = $129.60
				Total this account
03311.10	Concrete, footings, place	120	cy	**Labor:** 1 Foreman @ $17.20 = $17.20 1 Mason @ $14.40 = $14.40 4 Laborers @ $11.60 = $46.40 1 Operator @ $15.95 = $15.95 1 Oiler @ $12.08 = $12.08 1 Carpenter @ $16.20 = $16.20 Crew hourly rate = $122.23 Production rate = 24 cy per hour 120 ÷ 24 x $ 122.23 = $ 611.15 **Equipment:** 50 ton crane @ $108.00 Vibrators & bucket @ $ 8.00 Equipment hourly rate =$116.00 120 ÷ 24 x $116.00 = $580 **Material:** Transit Mix @ $49.50/cy, 5% waste 120 x 1.05 x $49.50 = $6,237
				Total this account
Total Bid Item No. 5				

Estimator: GAS

Labor Cost		Equipment Cost	Material Cost	Total Cost
Direct	Indirect			
$666	$320			$986
			$85	$85
			$108	$108
$666	$320	$0	$193	$1,179
$346	$138			$484
			$130	$130
$346	$138	$0	$130	$614
$611	$232			$843
		$580		$580
			$6,237	$6,237
$611	$232	$580	$6,237	$7,660
$1,623	$690			
	$2,313	$580	$6,560	$9,453

SUMMARY SHEET

Job: Highway Bridge   Bid Item No. 7: Concrete, deck slab, 10 in.

Cost Code	Work Type	Quantity	Unit	Calculations
03157.30	Deck forms, place & strip	1,800	sf	**Labor:** Labor unit cost $1.68 per sf 1,800 x $1.68 = $3,024 crane operator and oiler for 2 hours 2 x ($15.95 + $12.08) = $56 (Approx. 70% for placing, 30% for stripping) **Equipment:** 50 ton crane 2 hours @ $108.00 = $216 **Materials:** Material unit cost $0.73 per sf 1,800 x $0.73 = $1,314
				Total this account
03311.30	Concrete deck, place & screed	200 56	sy cy	**Labor:** Labor unit cost = $14.64 per cy 56 x $14.64 = $819.84 **Equipment:** Equip. unit cost = $12.43 per cy 56 x $12.43 = $696 **Material:** Transitmix @ $49.50 per cy 56 cy + 5 % waste 56 x $49.50 x 1.05 = $2,911
				Total this account
03345.30	Concrete deck, finish	1,800	sf	**Labor:** Labor unit cost = $0.91 per sf 1,800 x $0.91 = $1,638 **Equipment:** 2 trowelling machines @ $22 per day 2 x $22.00 = $44 **Material:** Material unit cost = $0.19 per sf 1,800 x $0.19 = $342
				Total this account
03370.30	Concrete deck, curing	1,800	sf	**Labor:** Labor unit cost = $0.03 per sf 1,800 x $0.03 = $54 **Material:** Material unit cost $0.02 per sf 1,800 x $0.02 = $36
				Total this account
03251.10	Concrete deck, saw joints	60	lf	**Labor:** Labor unit cost = $1.75 per ft 60 x $1.75 = $105 **Equipment:** Concrete saw @ $87 per day 1x $87.00 = $87
				Total this account
Total Bid Item No. 7				

Estimator: GAS

| Labor Cost | | Equipment | Material | Total |
Direct	Indirect	Cost	Cost	Cost
$3,024	$1,330			$4,354
$56	$22			$78
		$216		$216
			$1,314	$1,314
$3,080	$1,352	$216	$1,314	$5,962
$820	$320			$1,140
		$696		$696
			$2,911	$2,911
$820	$320	$696	$2,911	$4,747
$1,638	$655			$2,293
		$44		$44
			$342	$342
$1,638	$655	$44	$342	$2,679
$54	$20			$74
			$36	$36
$54	$20	$0	$36	$110
$105	$42			$147
		$87		$87
$105	$42	$87	$0	$234
$5,697	**$2,389**			
	$8,086	**$1,043**	**$4,603**	**$13,732**

SUMMARY SHEET

Job: Highway Bridge   Estimator: GAS

Bid Item No. 8a: Reinforcing steel

Cost Code	Work Type	Quantity	Unit	Calculations	Labor Cost		Equipment Cost	Material Cost	Total Cost
					Direct	Indirect			
03200.10	Steel, reinforcing, place	90,000	lb	**Labor:** Labor unit cost $.15 per lb 90,000 x $0.15 = $13,500	$13,500	$5,805			$19,305
				Equipment: Equipment unit cost = $0.03 per lb 90,000 x $0.03 = $2,700			$2,700		$2,700
				Materials: 45 tons @ $460 per ton = $20,700				$20,700	$20,700
				Total this account	$13,500	$5,805	$2,700	$20,700	$42,705
					$13,500	**$5,805**	**$2,700**	**$20,700**	
						$19,305			

Total Bid Item No. 8a   **$13,500**   **$19,305**   **$2,700**   **$20,700**   **$42,705**

SUMMARY SHEET

Job: Highway Bridge   Estimator: GAS

Bid Item No. 8b: Reinforcing steel (Subcontracted)

Cost Code	Work Type	Quantity	Unit	Calculations	Labor Cost		Equipment Cost	Material Cost	Total Cost
					Direct	Indirect			
03200.10	Steel, reinforcing, place	90,000	lb	Subbid quotation = $40,275					$40,275
					$0	**$0**	**$0**	**$0**	

Total Bid Item No. 8b   **$0**   **$0**   **$0**   **$0**   **$40,275**

SUMMARY SHEET

Job: Highway Bridge

Bid Item No. 9: Steel, structural

Estimator: GAS

Cost Code	Work Type	Quantity	Unit	Calculations	Labor Cost Direct	Labor Cost Indirect	Equipment Cost	Material Cost	Total Cost
05120.00	Steel, structural, place	65,500	lb	**Labor:**					
				Labor unit cost = $40.00 per ton					
				32.75 x $40.00 = $1,310.00	$1,310	$537			$1,847
				2 days of crane operator & oiler					
				16 x ($15.95 + $12.08) = $448	$448	$179			$627
				Equipment:					
				16 hours of crane time					
				16 x $108.00 = $1,728			$1,728		$1,728
				Material:					
				32.75 tons @ $826 per ton					
				32.75 x $826.00 = $27,052				$27,052	$27,052
				Total this account	$1,758	$716	$1,728	$27,052	$31,254
					$1,758	$716			
						$2,474			
Total Bid Item No. 9							$1,728	$27,052	$31,254

SUMMARY SHEET

Job: Highway Bridge

Bid Item No. 10: Bearing plates

Estimator: GAS

Cost Code	Work Type	Quantity	Unit	Calculations	Labor Cost Direct	Labor Cost Indirect	Equipment Cost	Material Cost	Total Cost
05812.00	Bearing plates	3,200	lb	**Labor:** Labor unit cost $.21 per lb 3,200 x $ 0.21 = $672 4 hours crane operator & oiler 4 x ($15.95 + $12.08) = $112	$672	$276			$948
					$112	$45			$157
				Equipment: 4 hours crane time 4 x $108.00 = $432			$432		$432
				Materials: Price quotation = $2,140				$2,140	$2,140
				Total this account	$784	$321	$432	$2,140	$3,677
					$1,105				
Total Bid Item No. 10					$784	$321	$432	$2,140	$3,677

SUMMARY SHEET

Job: Highway Bridge   Estimator: GAS

Bid Item No. 11: Guardrails

Cost Code	Work Type	Quantity	Unit	Calculations	Labor Cost		Equipment Cost	Material Cost	Total Cost
					Direct	Indirect			
05520.00	Guardrails	120	ft	**Labor:**					
				Labor unit cost $6.25 per ft					
				120 x $6.25 = $750	$750	$300			$1,050
				4 hours crane operator & oiler					
				4 x ($15.95 + $12.08) = $112	$112	$45			$157
				Equipment:					
				4 hours crane time					
				4 x $108.00 = $432			$432		$432
				Materials:					
				Price quotation = $3,752				$3,752	$3,752
				Anchor bolts 64 ea. @ $ 1.50 = $96				$96	$96
				Total this account	$862	$345	$432	$3,848	$5,487
					$862	$345			
						$1,207	$432	$3,848	$5,487

Total Bid Item No. 11

SUMMARY SHEET

Job: Highway Bridge   Estimator: GAS

Bid Item No. 12: Painting

Cost Code	Work Type	Quantity	Unit	Calculations	Labor Cost		Equipment Cost	Material Cost	Total Cost
					Direct	Indirect			
	Painting	job	ls	Subbid quotation = $5,820				$5,820	$5,820
					$0	$0	$0	$0	
						$0	$0	$0	$5,820

Total Bid Item No. 12

SUMMARY SHEET

Job: Highway Bridge   Field Overhead: Move In & Clean Up

Cost Code	Work Type	Quantity	Unit	Calculations
01500	Move In	1	ls	**Labor:**
				1 Foreman @ $17.20 = $17.20
				2 Carpenters @ $16.20 = $32.40
				4 Laborers @ $11.60 = $46.40
				2 Operator @ $14.51 = $29.02
				4 Truck drivers @ $13.80 = $55.20
				Crew hourly rate = $180.22
				Three 8-hour days required
				24 x $180.22 = $4,325
				Equipment:
				50 ton crane @ $108.00 = $108.00
				25 ton crane @ $ 76.50 = $76.50
				4 Flat bed trucks @ $ 15.00 = $60.00
				Equipment hourly rate =$244.50
				12 hours required
				12 x $244.50 = $2,934
				Total this account
01700	Clean Up	1	ls	**Labor:**
				1 Foreman @ $17.20 = $17.20
				6 Laborers @ $11.60 = $69.60
				Crew hourly rate = $86.80
				Two 8-hour days required
				16 x $86.80 = $1,389
				1 Foreman @ $17.20 = $17.20
				3 Laborers @ $11.60 = $34.80
				1 Operator @ $14.51 = $14.51
				2 Truck drivers @ $13.80 = $27.60
				Crew hourly rate = $ 94.11
				One 8-hour day required
				8 x $94.11 = $753
				Equipment:
				50 ton crane @ $108.00 = $108.00
				2 Flat bed trucks @ $ 15.00 = $30.00
				Equipment hourly rate =$138.00
				One 8-hour day required
				8 x $138.00 = $1,104
				Total this account

Total Move In & Clean Up

	Labor Cost		Equipment Cost	Material Cost	Total Cost
	Direct	Indirect			
					Estimator: GAS
	$4,325	$1,514			$5,839
			$2,934		$2,934
	$4,325	$1,514	$2,934	$0	$8,773
	$1,389	$430			$1,819
	$753	$264			$1,017
			$1,104		$1,104
	$2,142	$694	$1,104	$0	$3,940
	$6,467	$2,208			
		$8,675	$4,038	$0	$12,713

Highway Bridge Project Outline

I. Highway Bridge

A. Procurement
    1. Prepare & approve shop drawings for major materials
       a. Footing rebar
       b. Abutment & deck rebar
       c. Girders
    2. Fabricate and deliver major materials
       a. Footing rebar
       b. Abutment & deck rebar
       c. Girders
       d. Piles
B. Field mobilization & sitework
    1. Move in
    2. Prefabricate abutment forms
    3. Mobilize pile driving rig
    4. Excavation
       Excavate abutment #1
       Excavate abutment #2
    5. Demobilize pile driving rig
C. Pile foundations
    1. Pile driving
       a. Drive piles abutment #1
       b. Drive piles abutment #2

 2. Footing #1
   a. Forms & rebar footing #1
   b. Pour footing #1
   c. Strip footing #1
 3. Footing #2
   a. Forms & rebar footing #2
   b. Pour footing #2
   c. Strip footing #2
D. Abutments & wingwalls
 1. Abutment #1
   a. Forms & rebar abutment #1
   b. Pour abutment #1
   c. Strip & cure abutment #1
   d. Backfill abutment #1
   e. Rub concrete abutment #1
 2. Abutment #2
   a. Forms & rebar abutment #2
   b. Pour abutment #2
   c. Strip & cure abutment #2
   d. Backfill abutment #2
   e. Rub concrete abutment #2
E. Deck
 1. Set girders
 2. Deck Concrete operations
   a. Deck forms & rebar
   b. Pour & cure deck
   c. Strip deck
   d. Saw joints
F. Finishing operations
 1. Painting
 2. Guardrails
 3. Cleanup
 4. Final inspection

Appendix C

The PERT Procedure

In Section 5.1, a distinction was made between CPM and PERT. CPM is widely used in the construction industry while PERT has found its major application in research and development projects. This is largely due to the fact that PERT models the probabilistic nature of R & D work while CPM assumes sufficient past experience exists to estimate relatively accurate activity durations. In reality, even with an extensive experience base, the construction manager can learn several important lessons from a brief overview of the PERT procedure.

PERT assumes a high degree of variability in activity durations, making an accurate single estimate difficult to make. As a result, PERT requires three estimates of duration for each activity. The first is an optimistic estimate. This is the duration if everything goes very well with no problems. The probability of achieving an optimistic duration is about one chance in 100. The second estimate is a pessimistic estimate. This duration assumes that things go badly. This is the worst case scenario. The probability of completion within a pessimistic duration is about 99 chances in 100. The last estimate is the modal estimate. This duration is the best guess given the circumstances. The modal estimate is the one we normally use in CPM. I t is based on experience and tempered with current circumstances. It is important to remember that CPM uses a modal estimate because it will be shown later that a modal estimate is often a biased estimate.

In PERT, the three time estimates are designated t_o for the optimistic estimate, t_p for the pessimistic estimate, and t_m for the modal estimate. Figure C.1 shows that the estimated time durations of a PERT activity can be distributed any of three possible ways; normal, skewed to the right, or skewed to the left. Estimated values for t_o, t_p, and t_m define a given distribution. Often the activity distributions in CPM and PERT

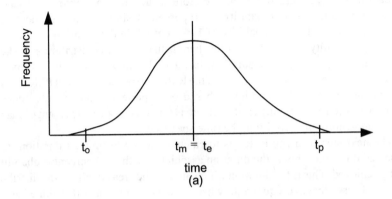

time
(a)

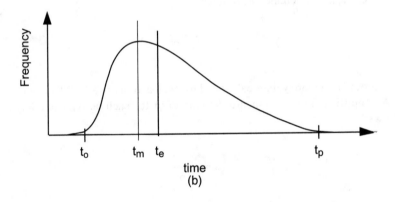

time
(b)

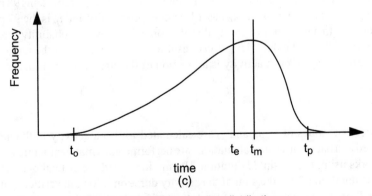

time
(c)

Figure C.1   *Activity duration distributions.*

networks are skewed to the right because it is the nature of projects that more things can go unexpectedly wrong than unexpectedly right.

In Figure C.1, the area under each of the distribution curves represents the probability of a particular duration. Therefore, a duration that divides the area under the curve in half, has a probability of 50 percent. Using this notion, the duration associated with t_o has a probability of 1 percent while the duration of t_p has a 99 percent probability. The point to notice here is that when the distribution is skewed to the right, t_m, the modal duration, has a probability of less than 50 percent, sometimes much less. Therefore, one of the lessons to be learned from PERT is: since the estimate of a modal duration is based on experience and recorded data, the duration is often biased and, therefore, there is often less than a 50 percent probability of accomplishing the activity in the time specified.

The next step in using PERT is to calculate a new estimated duration, t_e, the expected duration. This is the duration associated with a 50 percent probability of being achieved. This is the duration which divides the area equally in the distribution curve. Observe that t_e is equal to t_m when the curve is normally distributed as in (a) of Figure C.1. In (b) of Figure C.1, t_e is to the right of t_m and in (c) the opposite is true. Numerically, t_e is calculated as follows:

$$t_e = \frac{t_o 4t_m t_p}{6}$$

There are two more values calculated for each activity in a PERT network, s_t and v_t. A simplified form of standard deviation (s_t) for each activity is computed as follows:

$$s_t = \frac{t_p - t_o}{6}$$

The notion behind this evaluation of the standard deviation is that $t_p - t_o$ represents the range of durations for the activity, and moreover, this range represents 98 percent of the possible durations since t_o has a 1 percent probability and t_p is the 99 percent probability. In a normal probability distribution, a 98 percent probability is equal to approximately ±3 standard deviations, resulting in the division of the range by 6. The variance (v_t) of each activity is a function of the standard deviation.

$$v_t = s_t^2$$

Once the values of t_e, s_t, and v_t are calculated for each activity in the network, forward and backward pass calculations are performed in a manner identical to CPM networks using t_e as a single duration. Having identified the critical path, the total project duration (T_e) is the sum of the activity durations along the critical path. The value T_e is the expected value of the total project duration. Similar to the activity durations, T_e has an associated probability of 50 percent. Unlike the activity durations, T_e has a normal probability distribution regardless of the distributions of the individual activities.

The lesson here for CPM practitioners is that the estimated total project duration in CPM as well as PERT has an associated probability of only 50 percent. Construction contracts are usually quite specific regarding construction completion and a 50 percent probability of completion is typically inappropriate. To increase the probability of achieving completion by a promised date, extra days are added in the form of a contingency. This extra contingency time is not to be regarded as padding the schedule but rather a recognition that individual activity durations tend to be biased on the low side and collectively they underestimate the total project duration.

Using PERT, the amount of additional project time necessary to increase the probability of total project completion to an acceptable level is a function of the shape of the T_e normal distribution curve. If the curve is high and narrow, then the range of completion times will be correspondingly narrow. If the curve is low and wide, this reflects a great deal of uncertainty and a wide range of possible completion dates. The shape of the curve is determined from the sum of the variances (v_t) of each of the activities in the critical path.

$$V_t = \sum_{i=1}^{n} v_{ti}$$

The standard deviation for the project S_t is the square root of V_t.

$$S_t = \sqrt{V_t}$$

A simple example is in order. Figure C.2 is a PERT network showing ES, EF, LS, and LF dates based on the values of t_e for each activity. An example of the computations for t_e and v_t for activity A are shown below.

$$t_e = \frac{3 + 4 \times 12 + 21}{6} = 8$$

$$v_t = \frac{21 - 3}{6} = 3$$

Values for all of the other activities in the network are calculated in a similar manner.

Forward and backward pass calculations using values of t_e as durations, determine a critical path through activities A, C, and E. The estimated project duration is 37 days with a probability of 50 percent. The variance and standard deviation of the critical path are shown below.

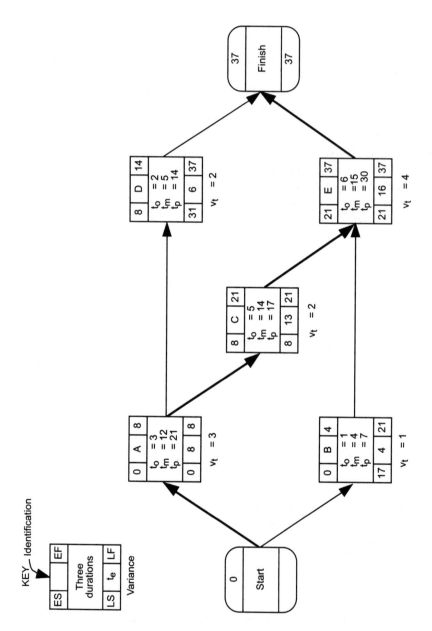

Figure C.2 Example PERT network.

Activity	t_e	v_t	
A	8	3	
C	13	2	
E	16	4	
Totals	$T_e = 37$	$V_t = 9$	$S_t = \pm 3$ days

With a total estimated total project duration of 37 days and a project standard deviation of ± 3 days, it is possible to make additional probability statements about the project. Figure C.3 is a normal distribution curve for the entire project. Note that the center of the curve is set at 37 days and the width of the curve is determined by S_t of ± 3 days. This figure also shows that the area of the curve between ± 1 standard deviation represents 68 percent of the distribution. This percentage can be confirmed by integrating the normal distribution function from $-1S_t$ to $+1S_t$ or by referring to a table of normal probabilities. Similarly, the area covered by ± 2 standard deviations represents 95 percent of the distribution.

The probability of finishing the project in 37 days is 50 percent. By moving forward one standard deviation (3 days) to 40 days, the area of the curve from the origin to 40 days is 84 percent (50 percent plus half of 68 percent). Therefore, the probability of finishing the project within 40 days is 84 percent. By the same reasoning, the probability of finishing the project within 44 days is 97.5 percent (50 percent plus half of 95 percent).

CPM procedures do not provide data to determine a probability distribution curve like the one in Figure C.3, but the principles are exactly the same. The project duration resulting from a forward and backward pass calculation has a 50 percent probability. To increase the probability to an acceptable level, extra time must be added. The amount of time is a function of the skewed variability of the activities and must be determined from experience.

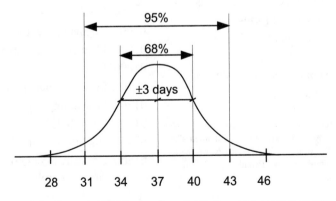

Figure C.3 *Example PERT network, estimated project duration distribution.*

Appendix D

Masterformat, Broadscope Section Titles*

Bidding Requirements, Contract Forms, and Conditions of the Contract

00010 Pre-bid information
00100 Instructions to bidders
00200 Information available to bidders
00300 Bid forms
00400 Supplements to bid forms
00500 Agreement forms
00600 Bonds and certificates
00700 General conditions
00800 Supplementary conditions
00900 Addenda

Note: The items listed above are not specification sections and are referred to as "Documents" rather than "Sections" in the Master List of Section Titles, Numbers, and Broadscope Section Explanations.

Specifications

Division 1—General Requirements

01010 Summary of work
01020 Allowances
01025 Measurement and payment
01030 Alternates/alternatives
01035 Modification procedures
01040 Coordination
01050 Field engineering
01060 Regulatory requirements
01070 Identification systems
01090 References
01100 Special project procedures
01200 Project meetings
01300 Submittals
01400 Quality control
01500 Construction facilities and temporary controls

---

* This listing has been reproduced with the permission of the Construction Specifications Institute, 601 Madison Street, Alexandria, Virginia, 22314-1791.

01600 Material and equipment

01650 Facility startup/ commissioning

01700 Contract closeout

01800 Maintenance

Division 2—Sitework

02010 Subsurface investigation

02050 Demolition

02100 Site preparation

02140 Dewatering

02150 Shoring and underpinning

02160 Excavation support systems

02170 Cofferdams

02200 Earthwork

02300 Tunneling

02350 Piles and caissons

02450 Railroad work

02480 Marine work

02500 Paving and surfacing

02600 Utility piping materials

02660 Water distribution

02680 Fuel and steam distribution

02700 Sewerage and drainage

02760 Restoration of underground pipe

02770 Ponds and reservoirs

02780 Power and communications

02800 Site improvements

02900 Landscaping

Division 3—Concrete

03100 Concrete framework

03200 Concrete reinforcement

03250 Concrete accessories

03300 Cast-in-place concrete

03370 Concrete curing

03400 Precast concrete

03500 Cementitious decks and toppings

03600 Grout

03700 Concrete restoration and cleaning

03800 Mass concrete

Division 4—Masonry

04100 Mortar and masonry grout

04150 Masonry accessories

04200 Unit masonry

04400 Stone

04500 Masonry restoration and cleaning

04550 Refractories

04600 Corrosion resistant masonry

04700 Simulated masonry

Division 5—Metals

05010 Metal materials

05030 Metal coatings

05050 Metal fastening

05100 Structural metal framing

05200 Metal joists

05300 Metal decking

05400 Cold formed metal framing

05500 Metal fabrications

05580 Sheet metal fabrications

05700 Ornamental metal

05800 Expansion control

05900 Hydraulic structures

Division 6—Wood and Plastics

06050 Fasteners and adhesives

06100 Rough carpentry

06130 Heavy timber construction

06150 Wood and metal systems

06170 Prefabricated structural wood

06200 Finish carpentry

06300 Wood treatment

06400 Architectural woodwork

06500 Structural plastics
06600 Plastic fabrications
06650 Solid polymer fabrications

*Division 7—Thermal and Moisture
Protection*
07100 Waterproofing
07150 Dampproofing
07180 Water repellents
07190 Vapor retarders
07195 Air barriers
07200 Insulation
07240 Exterior insulation and
finish systems
07250 Fireproofing
07270 Firestopping
07300 Shingles and roofing tiles
07400 Manufactured roofing and
siding
07480 Exterior wall assemblies
07500 Membrane roofing
07570 Traffic coatings
07600 Flashing and sheet metal
07700 Roof specialties and accessories
07800 Skylights
07900 Joint sealers

Division 8—Doors and Windows
08100 Metal doors and frames
08200 Wood and plastic doors
08250 Door opening assemblies
08300 Special doors
08400 Entrances and storefronts
08500 Metal windows
08600 Wood and plastic windows
08650 Special windows
08700 Hardware
08800 Glazing
08900 Glazed curtain walls

Division 9—Finishes
09100 Metal support systems
09200 Lath and plaster
09250 Gypsum board
09300 Tile
09400 Terrazzo
09450 Stone facing
09500 Acoustical treatment
09540 Special wall surfaces
09545 Special ceiling surfaces
09550 Wood flooring
09600 Stone flooring
09630 Unit masonry flooring
09650 Resilient flooring
09680 Carpet
09700 Special flooring
09780 Floor treatment
09800 Special coatings
09900 Painting
09950 Wall coverings

Division 10—Specialties
10100 Visual display boards
10150 Compartments and cubicles
10200 Louvers and vents
10240 Grilles and screens
10250 Service wall systems
10260 Wall and corner guards
10270 Access flooring
10290 Pest control
10300 Fireplaces and stoves
10340 Manufactured exterior
specialties
10350 Flagpoles
10400 Identifying devices
10450 Pedestrian control devices
10500 Lockers
10520 Fire protection specialties
10530 Protective covers

10550 Postal specialties
10600 Partitions
10650 Operable partitions
10670 Storage shelving
10700 Exterior protection devices
for openings
10750 Telephone specialties
10800 Toilet and bath accessories
10880 Scales
10900 Wardrobe and closet
specialties

Division 11—Equipment
11010 Maintenance equipment
11020 Security and vault equipment
11030 Teller and service equipment
11040 Ecclesiastical equipment
11050 Library equipment
11060 Theater and stage equipment
11070 Instrumental equipment
11080 Registration equipment
11090 Checkroom equipment
11100 Mercantile equipment
11110 Commercial laundry and dry
cleaning equipment
11120 Vending equipment
11130 Audio-visual equipment
11140 Vehicle service equipment
11150 Parking control equipment
11160 Loading dock equipment
11170 Solid waste handling
equipment
11190 Detention equipment
11200 Water supply and treatment
equipment
11280 Hydraulic gates and valves
11300 Fluid waste treatment and
disposal equipment
11400 Food service equipment

11450 Residential equipment
11460 Unit kitchens
11470 Darkroom equipment
11480 Athletic, recreational, and
therapeutic equipment
11500 Industrial and process
equipment
11600 Laboratory equipment
11650 Planetarium equipment
11660 Observatory equipment
11680 Office equipment
11700 Medical equipment
11780 Mortuary equipment
11850 Navigation equipment
11870 Agricultural equipment

Division 12—Furnishings
12050 Fabrics
12100 Artwork
12300 Manufactured casework
12500 Window treatment
12600 Furniture and accessories
12670 Rugs and mats
12700 Multiple seating
12800 Interior plants and planters

Division 13—Special Construction
13010 Air supported structures
13020 Integrated assemblies
13030 Special purpose rooms
13080 Sound, vibration, and
seismic control
13090 Radiation protection
13100 Nuclear reactors
13120 Pre-engineered structures
13150 Aquatic facilities
13175 Ice rinks
13180 Site constructed incinerators
13185 Kennels and animal shelters

13200 Liquid and gas storage tanks

13220 Filter underdrains and media

13230 Digester covers and appurtenances

13240 Oxygenation systems

13260 Sludge conditioning systems

13300 Utility control systems

13400 Industrial and process control systems

13500 Recording instrumentation

13550 Transportation control instrumentation

13600 Solar energy systems

13700 Wind energy systems

13750 Cogeneration systems

13800 Building automation systems

13900 Fire suppression and supervisory systems

13950 Special security construction

Division 14—Conveying systems

14100 Dumbwaiters

14200 Elevators

14300 Escalators and moving walks

14400 Lifts

14500 Material handling systems

14600 Hoists and cranes

14700 Turntables

14800 Scaffolding

14900 Transportation systems

Division 15—Mechanical

15050 Basic mechanical materials and methods

15250 Mechanical insulation

15300 Fire protection

15400 Plumbing

15500 Heating, ventilating, and air conditioning

15550 Heat generation

15650 Refrigeration

15750 Heat transfer

15850 Air handling

15880 Air distribution

15950 Controls

15990 Testing, adjusting, and balancing

Division 16—Electrical

16050 Basic electrical materials and methods

16200 Power generation–built-up systems

16300 Medium voltage distribution

16400 Service and distribution

16500 Lighting

16600 Special systems

16700 Communications

16850 Electrical resistance heating

16900 Controls

16950 Testing

Appendix E

Analysis of Estimating Accuracy

E.1 OVERALL ESTIMATING ACCURACY

The following numerical example is presented to illustrate how a construction company can analyze the overall accuracy of its cost estimating. When making such a check, comparisons are made between estimated and actual total costs of completed projects. Only relatively recent projects should be considered because estimating personnel and pricing behavior have a tendency to change with time. Enough projects should be included to provide a statistically significant sample. A minimum number of something like 20 past projects might be required with larger samples providing somewhat more reliable results.

Suppose that a construction contracting firm has completed 68 contracts during the past five years. A ratio, R, is computed for each of these projects. R is equal to the actual total project cost divided by the estimated total cost. Cost in each instance includes all project expense, including field overhead, but not profit. These values of R are then grouped together into equal intervals, say 0.02 or 2 percent. Larger or smaller intervals can be used depending somewhat on the size of the sample chosen. In the example, assume that the distribution of R-values shown in columns 1 and 2 of Figure E.1 is obtained for the 68 completed jobs.

Using the data given in columns 1 and 2, the histogram in Figure E.2 is plotted. A frequency polygon is actually more useful and is drawn as dashed lines between the midpoints of the horizontal bars of the histogram. The frequency polygon is now smoothed by sketching in a smooth curve that closely approximates the linear segments of the polygon. The resulting curve affords a quick and nonquantitative grasp of the company's bidding accuracy over the past five years as will now be explained.

Intervals of R (1)	Frequency of Projects f_i (2)	Interval Midpoint R_i (3)	$f_i \times R_i$ (4)	Cumulative Frequency F (5)	Percentage Cumulative Frequency $\frac{F}{n} \times 100$ (6)	Interval Variance $(R_i - \overline{R})^2 \times f_i$ (7)
R<0.88	0	–	0.00	0	0	0.0000
0.88<R<0.90	1	0.89	0.89	1	1.47	0.0113
0.90<R<0.92	0	0.91	0.00	1	1.47	0.0000
0.92<R<0.94	3	0.93	2.79	4	5.88	0.0133
0.94<R<0.96	6	0.95	5.70	10	14.71	0.0130
0.96<R<0.98	12	0.97	11.64	22	32.35	0.0084
0.98<R<1.00	16	0.99	15.84	38	55.88	0.0007
1.00<R<1.02	14	1.01	14.14	52	76.47	0.0026
1.02<R<1.04	9	1.03	9.27	61	89.71	0.0101
1.04<R<1.06	3	1.05	3.15	64	94.12	0.0086
1.06<R<1.08	2	1.07	2.14	66	97.06	0.0108
1.08<R<1.10	1	1.09	1.09	67	98.53	0.0087
1.10<R<1.12	1	1.11	1.11	68	100.00	0.0129
1.12<R	0	–	0.00	68	100.00	0.0000
Totals	n = 68		67.76			0.1004

$$R = \frac{\text{Actual Project Cost}}{\text{Estimated Project Cost}}$$

$$s = \text{Standard Deviation} = \sqrt{\frac{\Sigma(R_i - \overline{R})^2 \times f_i}{n-1}}$$

$$\text{Average Value of } R = \overline{R} = \frac{67.76}{68} = 0.9965$$

$$s = \sqrt{\frac{0.1004}{68-1}} = 0.039$$

Figure E.1 *Analysis of R values.*

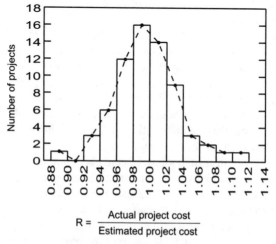

$$R = \frac{\text{Actual project cost}}{\text{Estimated project cost}}$$

Figure E.2 *Histogram of R values.*

It is reasonable to assume that estimating errors should be essentially random in nature and conform to the well-known normal distribution. Another way of saying this, perhaps, is that there should be an equal tendency to overprice and underprice a construction estimate. Consequently, it is suggested that a good estimating system is one whose ratios, R, are normally distributed with a mean value of 1.0 and a small range of maximum and minimum values. If the smoothed polygon obtained from the histogram of R-values resembles a normal distribution of small variance and with its mode at or very near a value of 1.0, such as that shown by curve A of Figure E.3, the overall bidding accuracy of the contractor has been good. Although curve B has its mode at 1.0, the large dispersion of values indicates loose estimating with a serious tendency to both underestimate and overestimate. A biased distribution such as curve C is indicative of systematic estimating errors or procedures that result in a pronounced tendency to consistently either underestimate or overestimate a majority of the jobs bid. If a smoothed polygon were to be sketched onto Figure E.2, the resulting curve would closely resemble a normal distribution with its mode at or very near a value, R, of 1.0.

Another plot of interest is cumulative frequency versus values of R. The values of cumulative frequency are computed in Figure E.1 and are plotted in Figure E.4. Figure E.4 is actually a "less-than" curve. To illustrate, this figure discloses that 38 of the 68 projects had an R value of less than 1.00. This means that on 38 of the completed projects, the actual costs were less than those estimated.

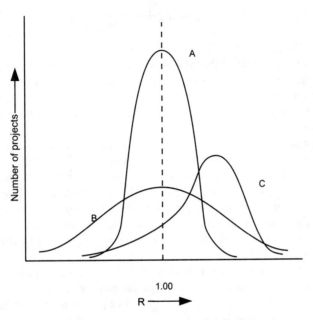

Figure E.3   *Bidding accuracy distribution.*

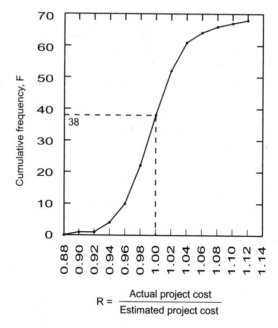

$$R = \frac{\text{Actual project cost}}{\text{Estimated project cost}}$$

Figure E.4 *Cumulative frequency of R values.*

E.2   EVALUATION OF ACCURACY

With regard to the evaluation of estimating accuracy, there are two measures of interest. one is the average or mean value of R. In this case, the average can be computed with satisfactory accuracy by the procedure shown in Figure E.1. As can be seen, the average value of the ratio of actual costs to estimated costs over the past five years has been very close to the value of 1.0. The significance of this is that, despite the fact that more jobs were overestimated than underestimated (38–30), there has been no consistent tendency to estimate too high or too low insofar as R-values are concerned.

The other measure of accuracy has to do with the dispersion or scatter of values of R. In our case, there were no values of R less than 0.88 nor larger than 1.12. This indicates, in a general fashion, that during the past years, projects have been both underestimated and overestimated by as much as approximately 12 percent. These values are indicative of extreme range, however, and do not tell the entire story. A much better measure of dispersion is the standard deviation that is computed in Figure E.1 and has a value of 0.039. The significance of this value will be discussed subsequently.

Another way discrete data can be checked for normal distribution is to plot them on normal probability paper. This is a special graph paper available at technical bookstores and similar outlets. Figure E.5 shows a plot on normal probability paper of percentage cumulative frequency verus the ratio, R. The values of percentage

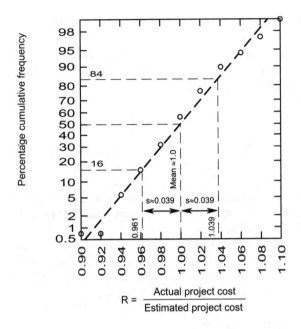

Figure E.5   *Normal probability plot.*

cumulative frequency have been calculated in Figure E.1. The closer the points plot to a straight line, the better the normality of a set of data. R-values deviating from a straight line indicate where the distribution is departing from normal. Figure E.5 shows that the distribution of R-values is close to normal except for the two extreme end points. The standard deviation of the R-values is found by the dashed line construction shown in Figure E.5, starting from the ordinate values of 84 and 16. The mean value of R is found from the ordinate value of 50. The data plot in Figure E.5 shows a mean value of very close to 1.00 and a standard deviation of 0.039, the same values as previously calculated in Figure E.1.

Having established that the R-values are very close to being normally distributed, the following observations can be made concerning the company's long-term bidding accuracy using well-known properties of the normal distribution curve. First, about 68 percent of the projects bid lie within plus or minus a standard deviation from the mean. Having established that the mean is essentially 1.0, this means that approximately 68 percent of the projects have a ratio, R, of between 0.961 and 1.039. Another way of saying this is that about two-thirds of the projects have a bidding error no larger than 3.9 percent of the total project cost. In this regard, the error can be either plus or minus. It is also true that about 95 percent of the projects are located within plus or minus two standard deviations from the mean. This says that the maximum bidding error for 95 percent of the jobs bid is no larger than plus or minus 7.8 percent. What constitutes acceptable values of standard deviation is now a matter for management decision.

E.3 ESTIMATING ACCURACY OF PROJECT COMPONENTS

The foregoing discussion of estimating accuracy presented in Appendix E has been based upon a comparative analysis of actual and estimated total project costs. Although a study of this kind provides the contractor with very valuable insight into the reliability of its overall estimating procedures, it does not provide information regarding the estimating accuracy of job components such as labor, equipment, materials, and project overhead.

Figure E.6 presents the results of a detailed estimating accuracy study made of ten construction projects that our contractor estimated and constructed during the year of 1990. This figure presents for each project, the R-values for labor, equipment, materials, and project overhead. In each case, R-values are obtained by dividing actual cost by estimated cost. Thus, R-values greater than 1.0 indicate underestimating and less than 1.0 indicate overestimating.

The R-values contained in Figure E.6 give the contractor a detailed and reliable indication of its recent estimating performance. Inspection of the data contained in Figure E.6 discloses that this contractor is more apt to underestimate labor and project overhead and overestimate equipment and materials.

Estimating Accuracy

Estimated Costs -->				
Job #	Labor	Equipment	Material	Overhead
9001	$242,146	$57,028	$237,922	$41,338
9002	$186,184	$78,867	$370,277	$64,066
9003	$12,463	$3,884	$26,018	$11,825
9004	$144,128	$71,165	$52,934	$50,751
9005	$278,934	$21,034	$310,225	$80,288
9006	$577,590	$194,280	$605,308	$104,596
9007	$52,534	$36,950	$76,541	$41,080
9008	$182,890	$86,416	$215,880	$66,014
9009	$161,492	$71,074	$220,315	$71,949
9010	$452,127	$129,532	$415,658	$112,371
Total	**$2,290,488**	**$750,230**	**$2,531,078**	**$644,278**

Actual Costs -->				
Job #	Labor	Equipment	Material	Overhead
9001	$258,975	$56,030	$228,643	$38,986
9002	$192,812	$78,173	$361,279	$62,022
9003	$12,394	$3,794	$24,683	$12,824
9004	$151,118	$71,983	$51,510	$52,502
9005	$286,577	$20,923	$306,037	$79,622
9006	$586,600	$195,601	$582,367	$100,600
9007	$52,376	$36,780	$75,500	$40,550
9008	$185,780	$85,336	$212,966	$67,922
9009	$161,169	$71,458	$219,456	$76,129
9010	$466,143	$128,677	$404,477	$111,382
Total	**$2,353,945**	**$748,755**	**$2,466,918**	**$642,540**

R Values -->				
Job #	Labor	Equipment	Material	Overhead
9001	1.070	0.983	0.961	0.943
9002	1.036	0.991	0.976	0.968
9003	0.995	0.977	0.949	1.085
9004	1.049	1.012	0.973	1.035
9005	1.027	0.995	0.987	0.992
9006	1.016	1.007	0.962	0.962
9007	0.997	0.995	0.986	0.987
9008	1.016	0.988	0.987	1.029
9009	0.998	1.005	0.996	1.058
9010	1.031	0.993	0.973	0.991
Average R	**1.023**	**0.995**	**0.975**	**1.005**
Std. deviation	**0.024**	**0.011**	**0.015**	**0.045**

Figure E.6   *Estimating accuracy of project components.*

Index